MASTERING WORKPLACE SKILLS: MATH FUNDAMENTALS

Related Titles

Mastering Workplace Skills: Grammar Fundamentals

Mastering Workplace Skills: Writing Fundamentals

MASTERING WORKPLACE SKILLS: MATH FUNDAMENTALS

NEW YORK

Published in the United States by LearningExpress, LLC, New York.

Cataloging-in-Publication Data is on file with the Library of Congress.

Printed in the United States of America

9 8 7 6 5 4 3 2 1

First Edition

ISBN 978-1-61103-019-8

For information on LearningExpress, other LearningExpress products, or bulk sales, please write to us at:
80 Broad Street
4th Floor
New York, NY 10004

Or visit us at:
www.learningexpressllc.com

CONTENTS

CONTENTS

Contributor

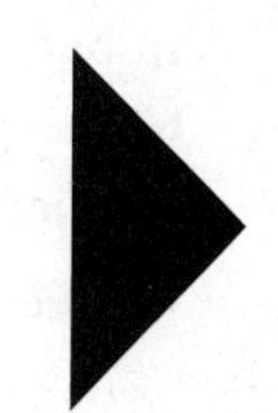

Mark A. McKibben is a professor of mathematics and computer science at Goucher College in Baltimore, Maryland. During his 12 years at this institution, he has taught more than 30 different courses spanning the mathematics curriculum, and has published two graduate-level books with CRC Press, more than two dozen journal articles about differential equations, and more than 20 supplements for undergraduate texts on algebra, trigonometry, statistics, and calculus.

Kimberly Stafford majored in mathematics and education at Colgate University in upstate New York. She taught math, science, and English in Japan, Virginia, and Oregon before settling in Los Angeles. Kimberly began her work in Southern California as an educator in the classroom but soon decided to launch her own private tutoring business so she could individualize her math instruction. She believes that a solid foundation in math empowers people by enabling them to make the best work, consumer, and personal decisions. Kimberly is unfazed by the ubiquitous student gripe, "When am I going to use this in real life?" She stresses that the mastery of math concepts that are less applicable to everyday life helps teach a critical skill—problem solving. The very ability to apply a set of tools to solve new and complex problems is an invaluable skill in both the workforce and personal life. Kimberly believes that mathematics is a beautiful arena for developing organized systems of thinking, clear and supported rationale, and effective problem solving.

Introduction

This book is for the *mathematically timid*—those who feel a flood of anxiety just looking at fractions and those who fall into a state of panic when having to deal with percentages. Well, you are not alone in your math phobia and since you've picked up this book, it's clear that you are ready to take some steps to leave that club and become mathematically competent. Congratulations! That is excellent news, because if you want to be a valuable employee in just about any field, you will need a solid foundation of math. Whether you will need to make change for a customer, order supplies while staying within a budget, or write a price estimate for a job your company is bidding on, a firm set of math skills will serve you well.

Mastering Workplace Skills: Math Fundamentals goes back to the basics, reteaching you the skills you've forgotten in a more approachable way. This time you will more deeply comprehend and retain what you learn. This book takes a fresh approach that uses examples and real-world illustrations while presenting mathematical skills in a user-friendly way. You will find yourself able to imagine being in the situations presented, which will help you grasp the material. In each lesson you will have the opportunity to practice applied problems that you might encounter in the workplace or in your personal life, using the skills taught in that lesson. It is very important to focus your energy on these problems because most math that we come across in our everyday lives does not look like a neat arithmetic problem on a math test!

Overcoming Math Anxiety

Do you love math? Do you hate math? Why? Stop right here, get out a piece of paper, and write the answers to these questions. Try to come up with specific reasons why you either like or don't like math. For instance, you may like math because you can check your answers and be sure they are correct. Or you may dislike math because it seems boring or complicated. Maybe you're one of those people who doesn't like math in a fuzzy sort of way but can't say exactly why. Now is the time to try to pinpoint your reasons. Figure out why you feel the way you do about math. If there are things you like about math and things you don't, write them down in two separate columns.

Once you get the reasons out in the open, you can address each one—especially the reasons you don't like math. You can find ways to turn those reasons into reasons you *could* like math. For instance, let's take a common complaint: Math problems are too complicated. If you think about this reason, you'll see that you can break every math problem down into small parts, or steps, and focus on one small step at a time. That way, the problem won't seem complicated. And, fortunately, all but the simplest math problems can be broken down into smaller steps.

If you're going to succeed at work, or just in your daily life, you're going to have to be able to deal with math. You need some basic math literacy to do well in many different kinds of careers. So if you have math anxiety or if you are mathematically challenged, the first step is to try to overcome your mental block about math. Start by remembering your past successes (Yes, everyone has them!). Then remember some of the nice things about math, things even a writer or artist can appreciate. Then, you'll be ready to tackle this book, which will make math as painless as possible.

Build on Past Success

Think back on the math you've already mastered. Whether or not you realize it, you already know a lot of math. For instance, if you give a cashier $20.00 for a book that costs $9.95, you know there's a problem if she only gives you $5.00 back. That's subtraction—a mathematical operation in action! Try to think of several more examples of how you unconsciously or automatically use your math knowledge.

Whatever you've succeeded at in math, focus on it. Perhaps you memorized most of the multiplication table and can spout off the answer to "What is 3 times 3?" in a second. Build on your successes with math, no matter how small they may seem to you now. If you can master simple math, then it's just a matter of time, practice, and study until you master more complicated math. Even if you have to revisit some lessons in this book to get the mathematical operations correct, it's worth it!

Great Things about Math

Math has many positive aspects that you may not have thought about before. Here are just a few:

1. Math is steady and reliable. You can count on mathematical operations to be constant every time you perform them: 2 plus 2 always equals 4. Math doesn't change from day to day depending on its mood. You can rely on each math fact you learn and feel confident that it will always be true.
2. If you work in fields such as the sciences, economics, nutrition, or business, you need math. Learning the basics

now will enable you to focus on more advanced mathematical problems and practical applications of math in these types of jobs.

3. Math is a helpful, practical tool that you can use in many different ways throughout your daily life, not just at work. For example, mastering the basic math skills in this book will help you complete practical tasks, such as balancing your checkbook, dividing your shared cell phone and data service plan between your family and friends, planning your retirement funds, or knowing the sale price of that new iPad that's been marked down 25%.
4. Mathematics is its own clear language. It doesn't have the confusing connotations or shades of meaning that sometimes occur in the English language. Math is a common language that is straightforward and understood by people all over the world.
5. Spending time learning new mathematical operations and concepts is good for your brain! You've probably heard this one before, but it's true. Working out math problems is good mental exercise that builds your problem-solving and reasoning skills. And that kind of increased brain power can help you in any field you want to explore.

These are just a few of the positive aspects of mathematics. Remind yourself of them as you work through this book. If you focus on how great math is and how much it will help you solve practical math problems in your daily life, your learning experience will go much more smoothly than if you keep telling yourself that math is terrible. Positive thinking really does work—whether it's an overall outlook on the world or a way of looking at a subject you're studying. Harboring a dislike for math could limit your achievement, so give yourself the powerful advantage of thinking positively about math.

How to Use This Book

Mastering Workplace Skills: Math Fundamentals is organized into small, manageable lessons—lessons you can master in a day. Each lesson presents a small part of a task one step at a time. The lessons teach by example—rather than by theory—so you have plenty of opportunities for successful learning. You'll learn by understanding, not by memorization.

Each new lesson is introduced with practical, easy-to-follow examples. Most lessons are reinforced by sample questions for you to try on your own, with clear, step-by-step solutions at the end of each lesson. You'll also find lots of valuable memory "hooks" and shortcuts to help you retain what you're learning. Practice question sets, scattered throughout each lesson, typically begin with easy questions to help build your confidence. As the lessons progress, easier questions are interspersed with the more challenging ones, so that even readers who are having trouble can successfully complete many of the questions. A little success goes a long way!

Exercises found in each lesson give you the chance to practice what you learned and apply each lesson's topic to real-world situations.

This book will get you ready to tackle math for work or for daily life by reviewing some of the math subjects you studied in grade school and high school, such as:

- **arithmetic:** fractions, decimals, percents, ratios and proportions, averages (mean, median, mode), probability, squares and square roots, length units, and word problems
- **elementary algebra:** positive and negative numbers, solving equations, and word problems
- **geometry:** lines, angles, triangles, rectangles, squares, parallelograms, circles, and word problems
- **probability:** determining the likelihood of certain events happening

This is a workbook, and as such, it's meant to be written in. Unless you checked it out from a library or borrowed it from a friend, write all over it! Get actively involved in doing each math problem—mark up the chapters boldly. You may even want to keep extra paper available, because sometimes you could end up using two or three pages of scratch paper for one problem—and that's fine!

Make a Commitment

You've got to take your math preparation further than simply reading this book. Improving your math skills takes time and effort on your part. You have to make the commitment. You have to carve time out of your busy schedule. You have to decide that improving your skills—improving your chances of doing well in almost any profession—is a priority for you.

If you're ready to make that commitment, this book will help you. Since each of its 18 lessons is designed to be completed in a day, you can build a firm math foundation in just one month, conscientiously working through the lessons five days a week. If you follow the tips for continuing to improve your skills and do each of the exercises, you'll build an even stronger foundation. Use this book to its fullest extent—as a self-teaching guide and then as a reference resource—to get the fullest benefit.

Now that you're armed with a positive math attitude, it's time to dig into the first lesson. Go for it!

CHAPTER 1

Working with Fractions

If I were again beginning my studies, I would follow the advice of Plato and start with mathematics.

—GALILEO GALILEI, mathematician and astronomer (1564–1642)

CHAPTER SUMMARY

This first fraction chapter will familiarize you with fractions, teaching you ways to think about them that will enable you to work with them more easily. This lesson introduces the three kinds of fractions and how to change from one kind of fraction to another, a useful skill for the fraction arithmetic that is covered in the upcoming chapters.

Fractions are one of the most important building blocks of mathematics and therefore understanding how to work with fractions is required in many different types of jobs. You encounter fractions every day: in recipes ($\frac{1}{2}$ cup of milk), driving ($\frac{3}{4}$ of a mile), measurements ($2\frac{1}{2}$ acres), money (half a dollar), and so forth. Most arithmetic problems involve fractions in one way or another. Decimals, percents, ratios, and proportions, which are covered in Chapters 6 through 12, are also fractions. To understand them, you must be very comfortable with fractions, which is what this lesson and the next four are all about.

What Is a Fraction?

A fraction is a representation of a part to a whole. When thinking of fractions, say "part to whole" to yourself and picture:

$$\frac{\text{Part}}{\text{Whole}} \rightarrow \frac{\text{\# of parts being represented}}{\text{\# of equal parts that make a whole}}$$

So if there is $\frac{3}{4}$ of a pan of brownies remaining, this means that a pan of brownies was divided into 4 equal parts, and 3 of them have not been eaten yet. The fraction $\frac{3}{4}$ is represented by the following illustration:

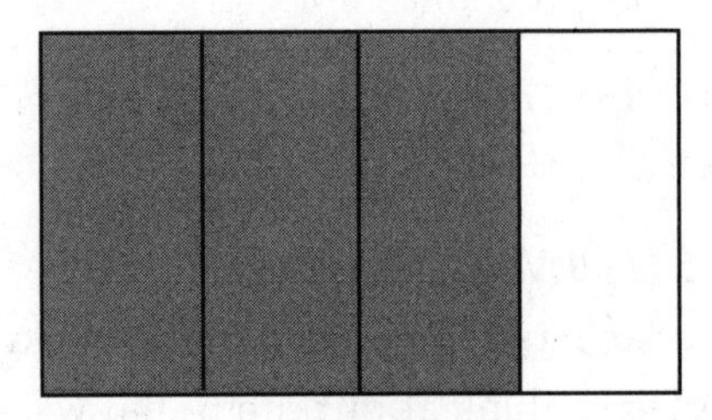

Now, let's look at the other ways fractions are used in everyday examples:

- **A minute is a fraction of an hour.** It is 1 of the 60 equal parts of an hour, or $\frac{1}{60}$ (one-sixtieth) of an hour.
- **The weekend days are a fraction of a week.** The weekend days are 2 of the 7 equal parts of the week, or $\frac{2}{7}$ (two-*sevenths*) of the week.
- **Coins are fractions of a dollar.** A nickel is $\frac{1}{20}$ (one-*twentieth*) of a dollar, because there are 20 nickels in one dollar. A dime

is $\frac{1}{10}$ (one-*tenth*) of a dollar, because there are 10 dimes in a dollar.

- **Measurements are expressed in fractions.** There are four quarts in a gallon. One quart is $\frac{1}{4}$ of a gallon. Three quarts are $\frac{3}{4}$ of a gallon.

TIP

It is important to understand what "0" represents in a fraction. Is $\frac{0}{5}$ the same as $\frac{5}{0}$? Definitely not! $\frac{0}{5} = 0$, because there are zero of five parts. But $\frac{5}{0}$ is undefined, because it is impossible to have five parts of zero. Zero is never allowed to be the denominator of a fraction!

The two numbers that compose a fraction are called the *numerator* and the *denominator*. The numerator is always in the top part of the fraction and the denominator is always in the bottom position:

$$\frac{\text{numerator}}{\text{denominator}}$$

For example, in the fraction $\frac{3}{8}$, the numerator is 3, and the denominator is 8. An easy way to remember which is which is to associate the word ***denominator*** with the word ***down***. The *denominator* represents the *total* number of equal parts that make up the whole, and the *numerator* represents the number of parts that are being considered. You can represent any fraction graphically by shading the number of parts being considered (numerator) out of the whole (denominator).

Example

Gina works at a pizza place where the pizzas are normally cut into 8 equal slices. Gina starts with a full pizza and sells 1 slice to Amara and 2 slices to Asad, so now 3 of the 8 slices are missing. The fraction $\frac{3}{8}$ represents the portion of the pizza that was sold. The following pizza shows this: It's divided into 8 equal slices, and 3 of the 8 slices (the ones sold) are shaded. Since the *whole* pizza was cut into 8 equal slices, 8 is the *denominator*. The part Gina sold was 3 slices, making 3 the *numerator*.

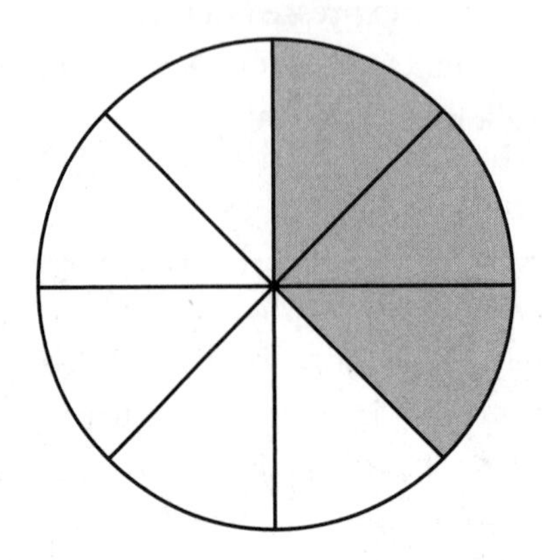

If you have difficulty conceptualizing a particular fraction, think in terms of *pizza fractions*. Just picture yourself eating the top number of slices from a pizza that's cut into the bottom number of slices. This may sound silly, but many people relate much better to visual images than to abstract ideas. Incidentally, this little trick comes in handy when comparing fractions to determine which one is bigger and when adding fractions.

Sometimes the *whole* isn't a *single* object like a pizza, but rather a *group* of objects. However, the shading idea works the same way. Four out of the following five triangles are shaded. Thus, $\frac{4}{5}$ of the triangles are shaded.

 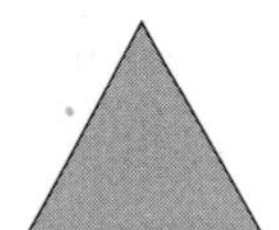   

Fractions and Money

A really important area where fractions can be applied is money. For example, suppose you have $5 that you are going to use to buy garlic knots from Gina at the pizza place. How many could you buy? Let's visualize the *whole* $5 broken into *parts* that are 50¢

each. Notice that each dollar bill is split into 2 groups of 50¢ each. How many groups of 50¢ does $5 result in?

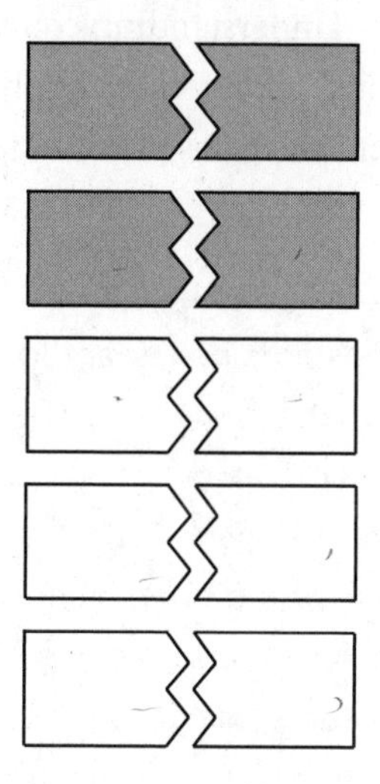

Since the whole $5 can be broken into 10, 50-cent parts, we conclude that you could buy 10 garlic knots.

A related question is: "What fraction of your $5 would be spent if you bought 4 garlic knots?" This can be visualized by shading 4 of the 10 parts, as shown in practice problem 1.

The fraction of the $5 is $\frac{4}{10}$.

Another way to consider buying 4 garlic knots is to recognize that this will cost you $2 out of your $5. This can be written in fraction form by using $2 as your *part* and $5 as your *whole*. This is represented as $\frac{2}{5}$ of your $5. You can see in the previous illustration that $2 out of the $5 became shaded when you bought 4 garlic knots for 50¢ each.

At first we looked at how 4 garlic knots for 50¢ each can be represented as using 4 out of your 10 parts of 50¢ each. Then we recognized that this same garlic knot purchase can also be represented as using 2 of your 5 one-dollar bills. Using fractions, these two different ways of framing the same purchase show that $\frac{4}{10}$ is the same thing as $\frac{2}{5}$. We can write this as $\frac{4}{10} = \frac{2}{5}$. When two fractions represent the same quantity, they are called *equivalent fractions.* The equivalent fraction with the smaller numbers is called a *reduced* fraction, so the $\frac{2}{5}$ is a reduced fraction. Understanding equivalent fractions is very important in order to add, subtract, multiply, and divide fractions, and we discuss equivalent fractions in more detail in later lessons.

When writing fractions, it is necessary for the numerator and denominator to be in the same units. For example, consider that Asriel has 3 feet of ribbon and she uses 7 inches of it to make a bow. It would be incorrect to say that she used $\frac{7}{3}$ of her ribbon since the 7 would be in inches and the 3 would be in feet. Instead, turn 3 feet into inches by multiplying it by the 12 inches that are in each foot: 3 ft. × 12 in. = 36 inches. Now we can say that Asriel used $\frac{7}{36}$ of her ribbon since both the numerator and denominator are in inches.

Practice

A fraction represents a *part* of a *whole.* Name the fractions that indicate the shaded parts. Answers are at the end of the lesson. If you are up for a challenge, see whether the fraction can be repreented in reduced form.

_____ **1.**

_____ **2.**

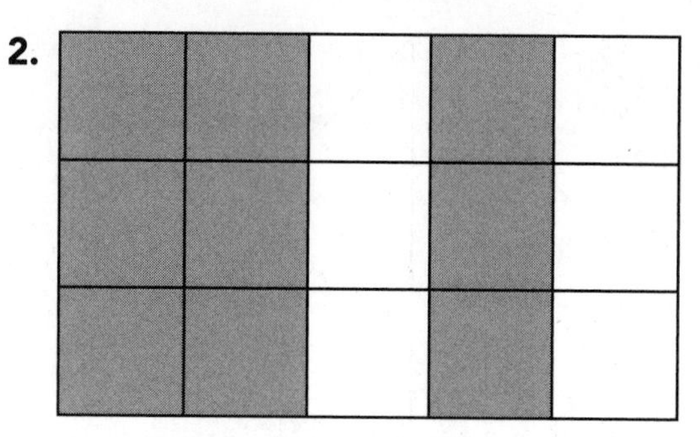

_____ **3.**

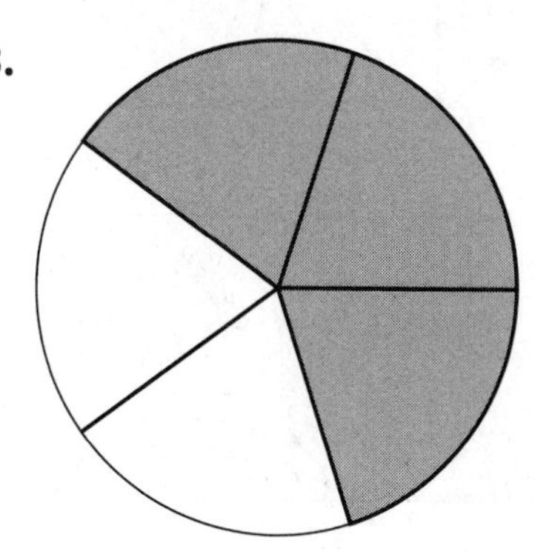

_____ **4.**

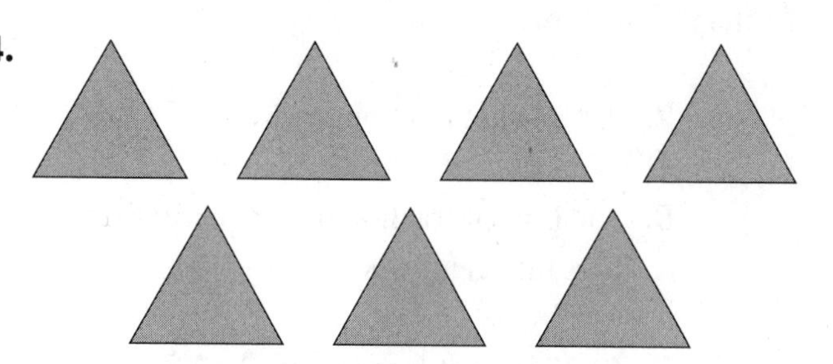

_____ **5.**

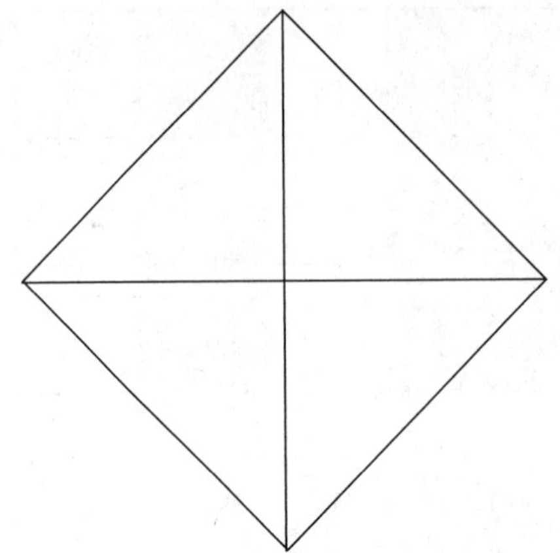

_____ **6.**

Money Problems

_____ **7.** 25¢ is what fraction of 75¢?

_____ **8.** 50¢ is what fraction of $1? (Hint: Convert both numbers into cents first.)

_____ **9.** $1.25 is what fraction of $10.00?

_____ **10.** $20.00 is what fraction of $200.00?

Distance Problems

Solve these distance problems by first converting both numbers into the same unit of measurement. Use these equivalents:

1 foot = 12 inches

1 yard = 3 feet = 36 inches

1 mile = 5,280 feet = 1,760 yards

_____ **11.** 8 inches is what fraction of a foot?

_____ **12.** 8 inches is what fraction of a yard?

_____ **13.** 1,320 feet is what fraction of a mile?

_____ **14.** 880 yards is what fraction of a mile?

Time Problems

Solve these time problems by first converting both numbers into the same unit of time. Use these equivalents:

1 minute = 60 seconds

1 hour = 60 minutes

1 day = 24 hours

_____ **15.** 20 seconds is what fraction of a minute?

_____ **16.** 3 minutes is what fraction of an hour?

_____ **17.** 30 seconds is what fraction of an hour?

_____ **18.** 80 minutes is what fraction of a day? (Hint: How many minutes are in a day?)

Three Kinds of Fractions

There are three kinds of fractions, each explained here.

Proper Fractions

In a *proper fraction*, the numerator is less than the denominator. Some examples are:

$$\frac{1}{2}, \frac{2}{3}, \frac{4}{9}, \frac{8}{13}$$

The value of a proper fraction is less than 1.

Example

Suppose you eat 3 slices of a pizza that's cut into 8 slices. Each slice is $\frac{1}{8}$ of the pizza. You've eaten $\frac{3}{8}$ of the pizza.

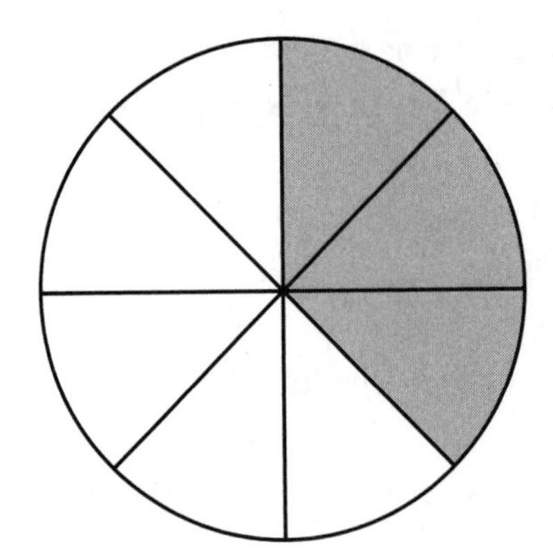

Improper Fractions

In an *improper fraction*, the numerator is greater than or equal to the denominator:

$$\frac{3}{2}, \frac{5}{3}, \frac{14}{9}, \frac{15}{12}$$

The value of an improper fraction is greater than or equal to 1.

- When the top and bottom numbers are the same, the value of the fraction is 1. For example, all of these fractions are equal to 1: $\frac{2}{2}$, $\frac{3}{3}$, $\frac{4}{4}$, $\frac{5}{5}$, etc.
- Any whole number can be written as an improper fraction by writing that number as the top number of a fraction whose bottom number is 1, for example, $\frac{4}{1} = 4$.

Example

Let's go back to Gina, who is working her shift at the pizza place during a busy lunch shift. Pete comes in and buys 4 slices for his friends. Then Celia comes in and buys 3 slices for her employees. Then Lauren gets 2 slices for herself. The first 8 slices Gina sold were $\frac{8}{8}$ of a pizza or 1 entire pie. The next slice sold brought Gina's lunch sales up to $\frac{9}{8}$. When Gina's boss asks her how much pizza she sold during lunch, he looks at her funny when she says $\frac{9}{8}$ pies. Then she rephrases it and tells him that she sold 1 and $\frac{1}{8}$ pizzas and he nods his head. Thank goodness Gina understands mixed numbers—let's take a closer look at those that follow so you can, too! But first, look at the following illustration to see what $\frac{9}{8}$ (or 1 and $\frac{1}{8}$), pizza looks like:

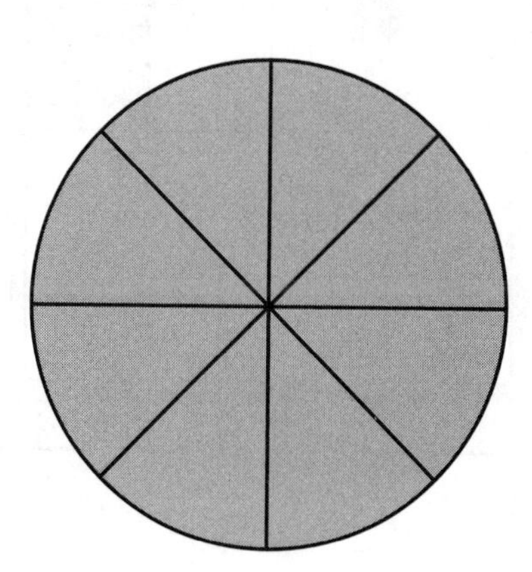

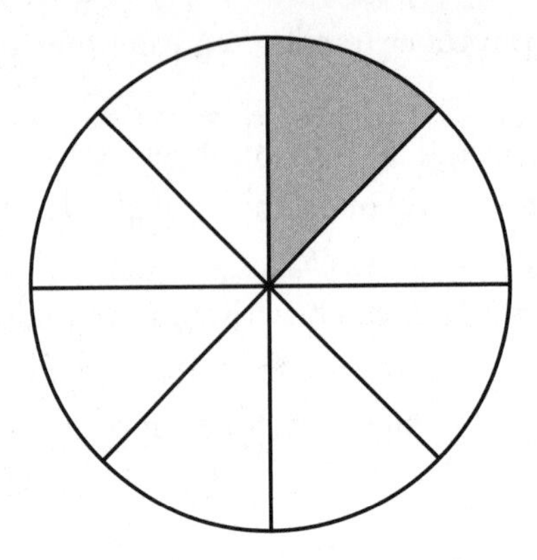

TIP

If a shape is divided into pieces of different sizes, you cannot just add up all the sections. Break the shape up into equal sections of the smaller pieces and use the total number of smaller pieces as your denominator. For example, break this box into 16 of the smaller squares instead of counting this as just six sections. The fraction that represents the shaded area would then be $\frac{3}{16}$.

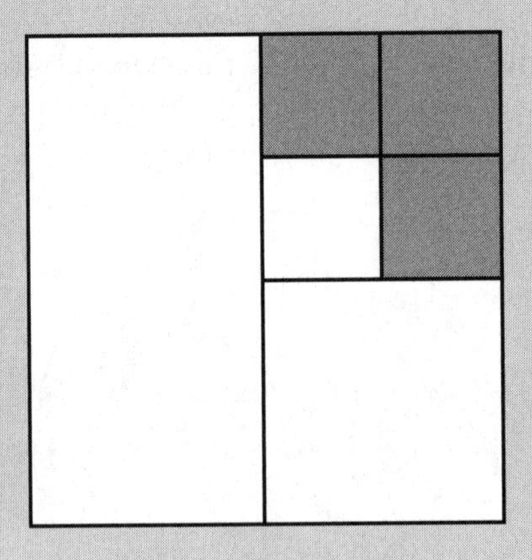

Mixed Numbers

When a proper fraction is written to the right of a whole number, the whole number and fraction together constitute a *mixed number*:

$$3\frac{1}{2}, 4\frac{2}{3}, 12\frac{3}{4}, 24\frac{2}{5}$$

The value of a mixed number is greater than 1: It is the sum of the whole number and a proper fraction.

It is convenient to clearly separate how many wholes from the remaining proper portion of a fraction when dealing with improper fractions.

Example

Remember those 9 slices that Gina sold during lunch? She first told her boss that she sold $\frac{9}{8}$ pies, but then she changed it to $1\frac{1}{8}$ pies after her boss was confused by her original answer. This means she sold 1 entire pie and $\frac{1}{8}$ of another pie.

⚠ CAUTION

It is important to realize that $\frac{1}{2}$ does not mean "6 times 12."

Changing Improper Fractions into Mixed or Whole Numbers

As Gina learned when talking to her boss, sometimes *mixed numbers* are easier to understand in context than *improper fractions.*

Therefore, it is important to be able to change improper fractions into mixed numbers by following these steps:

1. Divide the denominator into the numerator.
2. If there is a remainder, change it into a fraction by writing it as the remainder over the original denominator of the improper fraction. Write it next to the whole number.

Example
Change $\frac{13}{2}$ into a mixed number.

1. Divide the denominator (2) into the numerator (13) to get the whole number portion (6) of the mixed number: $2\overline{)13}$ = 6, 12, remainder 1
2. Write the remainder of the division (1) over the original denominator (2): $\frac{1}{2}$
3. Write the two numbers together: $6\frac{1}{2}$
4. Check: Change the mixed number back into an improper fraction (see steps starting on page 15). If you get the original improper fraction, your answer is correct.

Example
Change $\frac{12}{4}$ into a mixed number.

1. Divide the denominator (4) into the numerator (12) to get the whole number portion (3) of the mixed number: $4\overline{)12}$ = 3, 12, remainder 0
2. Since the remainder of the division is zero, you're done. The improper fraction $\frac{12}{4}$ is actually a whole number: 3
3. Check: Multiply 3 by the original bottom number (4) to make sure you get the original top number (12) as the answer.

Here is your first sample question in this book. Sample questions are a chance for you to practice the steps demonstrated in previous examples. Write down all the steps you follow when solving the question, and then compare your approach to the one demonstrated at the end of the lesson.

Sample Question 1

During the dinner rush, Gina sold $\frac{14}{3}$ pizzas. Her boss doesn't know what this means so she needs to change this into a mixed number for him. What will she report to her boss?

Practice

Help Gina change the following improper fractions of pizza sales to mixed numbers.

_____ **19.** $\frac{10}{3}$

_____ **20.** $\frac{15}{6}$

_____ **21.** $\frac{12}{7}$

_____ **22.** $\frac{6}{6}$

_____ **23.** $\frac{200}{25}$

_____ **24.** $\frac{75}{70}$

Changing Mixed Numbers into Improper Fractions

Since it is very complicated to multiply and divide numbers that are in mixed number format, it is important to know how to convert mixed numbers into improper fractions using the following steps:

1. Multiply the whole number by the denominator.
2. Add the product from step 1 to the numerator.
3. Write the total as the numerator of a fraction over the original denominator.

Example
Change $2\frac{3}{4}$ into an improper fraction.

1. Multiply the whole number (2) by the denominator (4): $2 \times 4 = 8$
2. Add the result (8) to the numerator (3): $8 + 3 = 11$
3. Put the total (11) over the denominator (4): $\frac{11}{4}$
4. Check: Reverse the process by changing the improper fraction into a mixed number. Since you get back $2\frac{3}{4}$, your answer is right.

Example
Change $3\frac{5}{8}$ into an improper fraction.

1. Multiply the whole number (3) by the denominator (8): $3 \times 8 = 24$
2. Add the result (24) to the numerator (5): $24 + 5 = 29$
3. Put the total (29) over the denominator (8): $\frac{29}{8}$
4. Check: Change the improper fraction into a mixed number. Since you get back $3\frac{5}{8}$, your answer is right.

Sample Question 2

Change $3\frac{2}{5}$ into an improper fraction.

Practice

Gina's boss entered last week's sales into the books as mixed numbers, but she prefers to have them as improper fractions. Help Gina change these mixed numbers into improper fractions.

_____ **25.** $1\frac{1}{2}$

_____ **26.** $2\frac{3}{8}$

_____ **27.** $7\frac{3}{4}$

_____ **28.** $10\frac{1}{10}$

_____ **29.** $15\frac{2}{3}$

_____ **30.** $12\frac{2}{5}$

TRY THIS

Do you have a place where you collect all your extra change? Maybe it's in your wallet, or in a dish at home, or in a compartment in your car. Gather all your spare change and add it up. Now express your total findings of change as a fraction over 100 (the 100 in the denominator represents the 100 pennies it takes to make a dollar). If you found more than $1 in change you should have an improper fraction. If so, change that improper fraction into a mixed number.

Answers

Practice Problems

1. $\frac{4}{10}$ or $\frac{2}{5}$
2. $\frac{9}{15}$ or $\frac{3}{5}$
3. $\frac{3}{5}$
4. $\frac{7}{7}$ or 1
5. $\frac{0}{4}$ or 0
6. $\frac{1}{1}$ or 1
7. $\frac{25}{75}$ or $\frac{1}{3}$
8. $\frac{50}{100}$ or $\frac{1}{2}$
9. $\frac{1.25}{10}$ or $\frac{1}{8}$
10. $\frac{20}{200}$ or $\frac{1}{10}$
11. $\frac{8}{12}$ or $\frac{2}{3}$
12. $\frac{8}{36}$ or $\frac{2}{9}$
13. $\frac{1,320}{5,280}$ or $\frac{1}{4}$
14. $\frac{880}{1,760}$ or $\frac{1}{2}$
15. $\frac{20}{60}$ or $\frac{1}{3}$
16. $\frac{6}{60}$ or $\frac{1}{20}$
17. $\frac{30}{3,600}$ or $\frac{1}{120}$
18. $\frac{80}{1,440}$ or $\frac{1}{18}$
19. $3\frac{1}{3}$
20. $2\frac{3}{6}$ or $2\frac{1}{2}$
21. $1\frac{5}{7}$
22. 1
23. 8
24. $1\frac{5}{70}$ or $1\frac{1}{14}$
25. $\frac{3}{2}$
26. $\frac{19}{8}$
27. $\frac{31}{4}$

28. $\frac{101}{10}$

29. $\frac{47}{3}$

30. $\frac{62}{5}$

Sample Question 1

1. Divide the denominator (3) into the numerator (14) to get the whole number portion (4) of the mixed number:

$$\begin{array}{r} 4 \\ 3\overline{)14} \\ \underline{12} \\ 2 \end{array}$$

2. Write the remainder of the division (2) over the original denominator (3): $\frac{2}{3}$
3. Write the two numbers together: $4\frac{2}{3}$
4. Check: Change the mixed number back into an improper fraction to make sure you get the original $\frac{14}{3}$.

Sample Question 2

1. Multiply the whole number (3) by the denominator (5): $3 \times 5 = 15$
2. Add the result (15) to the numerator (2): $15 + 2 = 17$
3. Put the total (17) over the denominator (5): $\frac{17}{5}$
4. Check: Change the improper fraction back to a mixed number.

$$\begin{array}{r} 3 \\ 5\overline{)17} \\ \underline{15} \end{array}$$

Dividing 17 by 5 gives an answer of 3 with a remainder of 2: 2

Put the remainder (2) over the original denominator (5): $\frac{2}{5}$

Write the two numbers together to get back the original mixed number: $3\frac{2}{5}$

CHAPTER

2 Converting Fractions

Do not worry about your difficulties in mathematics. I can assure you mine are still greater.

—ALBERT EINSTEIN, theoretical physicist (1879–1955)

CHAPTER SUMMARY

This chapter begins by showing you how to convert fractions into decimals. Then you'll learn how to reduce fractions and how to raise them to higher terms—skills you'll need to do arithmetic with fractions. Before actually beginning fraction arithmetic (which is in the next lesson), you'll learn some clever shortcuts for comparing fractions.

Chapter 1 defined a fraction as *a part of a whole*. This is definitely a convenient way to conceptualize fractions and to picture what they mean in real-world context. Oftentimes in the workplace, you will need to be able to translate fractions into another form that you may already be a little familiar with—decimals. Fractions are converted into decimal format to make them easier to compare and work with arithmetically. Here's an alternative way to think of fractions:

A fraction simply means "divide."
The top number of the fraction is divided
by the bottom number.

Thus, $\frac{3}{4}$ means "3 divided by 4," which may also be written as $3 \div 4$ or $4\overline{)3}$. The value of $\frac{3}{4}$ is the same as the *quotient* (result) you get when you perform the division. $3 \div 4 = 0.75$, which is the **decimal value** of the fraction. Notice that $\frac{3}{4}$ of a dollar (which is said "3 quarters") is the same thing as 75¢. 75¢ can also be written as $0.75, which is the decimal value of $\frac{3}{4}$.

Example
Find the decimal value of $\frac{4}{25}$.

Divide 25 into 4. (Note that you will need to add a decimal point and a series of zeros to the end of the 4.)

$$\begin{array}{r} 0.16 \\ 25\overline{)4.00} \\ \underline{-25} \\ 150 \\ \underline{-150} \\ 0 \end{array}$$

So, the fraction $\frac{4}{25}$ is equivalent to the decimal 0.16.

Example
Find the decimal value of $\frac{1}{9}$.

Divide 9 into 1 (note that you have to add a decimal point and a series of zeros to the end of the 1 in order to divide 9 into 1):

```
  0.1111 etc.
9)1.0000 etc.
  9
   10
    9
    10
     9
     10
```

The fraction $\frac{1}{9}$ is equivalent to the *repeating decimal* 0.1111 etc., which can be written as $0.\overline{1}$. (The little "hat" over the 1 indicates that it repeats indefinitely.)

The rules of arithmetic do not allow you to divide by zero. Thus, zero can never be the denominator of a fraction.

TIP

The following is a list of fractions that commonly occur. Notice the patterns of these fractions and begin to commit these fraction and decimal equivalences to memory:

- **$\frac{1}{3}$s are repeating decimals that increase by $0.3\overline{3}$:**
  - $\frac{1}{3} = 0.3\overline{3}$
  - $\frac{2}{3} = 0.66$
- **$\frac{1}{4}$s increase by 0.25:**
  - $\frac{1}{4} = 0.25$
  - $\frac{2}{4} = 0.50$
  - $\frac{3}{4} = 0.75$
- **$\frac{1}{5}$s increase by 0.2:**
  - $\frac{1}{5} = 0.2$
  - $\frac{2}{5} = 0.4$
  - $\frac{3}{5} = 0.6$
  - $\frac{4}{5} = 0.8$

Certain mathematical operations are easier to handle with fractions and others are easier to compute with decimals. One example where decimals are preferred is in the case of Eli, an intern for a landscape architect. Contractors submit measurements in fractions to Eli, but his boss needs the measurements to be in decimal form. Another person who needs to perform lots of fraction-decimal conversion is Dawn. She is an accountant who needs to turn interest rates into decimals so she can determine her client's monthly mortgage payments, since the rates are always given to her as fractions.

Practice

Help Eli and Dawn turn the following fractions into decimals.

_____ **1.** $\frac{1}{2}$

_____ **2.** $\frac{1}{4}$

_____ **3.** $\frac{3}{4}$

_____ **4.** $\frac{1}{3}$

_____ **5.** $\frac{2}{3}$

_____ **6.** $\frac{1}{8}$

_____ **7.** $\frac{3}{8}$

_____ **8.** $\frac{5}{8}$

_____ **9.** $\frac{7}{8}$

_____ **10.** $\frac{1}{5}$

_____ **11.** $\frac{2}{5}$

_____ **12.** $\frac{3}{5}$

_____ **13.** $\frac{4}{5}$

_____ **14.** $\frac{1}{10}$

The decimal values you just computed are worth memorizing. They are among the most common fraction-to-decimal equivalents you will encounter in the workplace.

Reducing a Fraction

Reducing a fraction means writing it in *lowest terms.* For instance, 50¢ is $\frac{50}{100}$ of a dollar, or $\frac{1}{2}$ of a dollar. In fact, if you have 50¢ in your pocket, you say that you have *half* a dollar. We say that the fraction $\frac{50}{100}$ *reduces* to $\frac{1}{2}$. Reducing a fraction does not change its value. **When you do arithmetic with fractions, always reduce your answer to lowest terms.** To reduce a fraction:

1. Look for the largest whole number factor that divides *evenly* into both the numerator and the denominator (you can use any number that divides both terms, but if you use a larger factor for the division you'll have less steps).
2. Divide the factor into the numerator and denominator and write the division answers in a new fraction.
3. Repeat steps 1 and 2 until there is no number that will evenly divide into both the numerator and the denominator. Now the fraction is in lowest terms.

Notice why it is faster to reduce when you find the largest number that divides evenly into both the top and bottom numbers of the fraction:

Example
Reduce $\frac{8}{24}$ to lowest terms.

Two steps:
1. Divide by 4: $\frac{8 \div 4}{24 \div 4} = \frac{2}{6}$
2. Divide by 2: $\frac{2 \div 2}{6 \div 2} = \frac{1}{3}$

One step:
1. Divide by 8: $\frac{8 \div 8}{24 \div 8} = \frac{1}{3}$

Now you try it. Solutions to sample questions are at the end of the chapter.

Sample Question 1

Katherine works at a fabric store. A customer would like to buy 30 inches of fabric that is priced by the yard. Katherine cannot write "$\frac{30}{36}$ of a yard" on the customer's slip because the store policy is to write all measurements in lowest terms. Help Katherine reduce $\frac{30}{36}$ to lowest terms.

Reducing Shortcut

When the numerator and denominator both end in zeros, cross out the same number of zeros in both numbers to begin the reducing process. (Crossing out zeros is the same as dividing by 10; 100; 1,000; etc., depending on the number of zeros you cross out.) For example, $\frac{300}{4{,}000}$ reduces to $\frac{3}{40}$ when you cross out two zeros in both numbers:

$$\frac{300 \div 100}{4{,}000 \div 100} = \frac{3\not{0}\not{0}}{4{,}0\not{0}\not{0}} = \frac{3}{40}$$

There are tricks to see if a number is divisible by 2, 3, 4, 5, and 6. Use the tricks in this table to find the best number to use when reducing fractions to lowest terms:

DIVISIBILITY TRICKS	
A # IS DIVISIBLE BY . . .	**. . . WHEN THE FOLLOWING IS TRUE.**
2	the number is even
3	the sum of all the digits is divisible by three
4	the last two digits are divisible by 4
5	the number ends in 0 or 5
6	the number is even and is divisible by 3

Practice

Help Katherine reduce the following fabric measurements into lowest terms.

_____ **15.** $\frac{8}{36}$

_____ **16.** $\frac{12}{72}$

_____ **17.** $\frac{72}{36}$

_____ **18.** $\frac{24}{36}$

Sierra works in lumberyard where a hardwood trim is sold in 84-inch lengths. Customers can have pieces custom cut and Sierra must represent the following cuts in lowest terms.

____ **19.** $\frac{60}{84}$

____ **20.** $\frac{48}{84}$

____ **21.** $\frac{12}{84}$

Practice the reducing shortcut that was given for multiples of 10:

____ **22.** $\frac{20}{700}$

____ **23.** $\frac{2,500}{5,000}$

____ **24.** $\frac{1,500}{75,000}$

Raising a Fraction to Higher Terms

The opposite of reducing a fraction is raising it to higher terms. It is often necessary to raise a fraction to higher terms in order to create common denominators. (Common denominators are needed to add and subtract fractions; we cover these skills in the next chapter.) To create an equivalent fraction that is in higher terms you simply need to multiply the numerator and denominator by the same number and rewrite the *products* (*product* means multiplication answer) as a new fraction. In this manner, $\frac{2}{3}$ can become $\frac{18}{27}$:

$$\frac{2}{3} = \frac{2 \times 9}{3 \times 9} = \frac{18}{27}$$

You will normally need to raise a fraction to higher terms with a specific denominator in mind. This skill is needed to add and subtract fractions as well as to convert fractions into percentages. To raise a fraction to higher terms with a specific denominator:

1. Divide the desired denominator by the original denominator to find the multiplication factor you will use to create the new fraction.
2. Multiply the factor found in step 1 by the numerator and denominator and write the products (multiplication answers) as a new fraction.
3. Check your answer by reducing it. You should arrive back at the original fraction.

Example:
Raise $\frac{5}{8}$ to 24ths

1. Divide the desired denominator (24) by the original denominator (8) to find the multiplication factor you will use to create the new fraction:

$$24 \div 8 = 3$$

2. Multiply the factor found in step 1 (3) by the numerator (5) and denominator (8) and write the products (multiplication answers) as a new fraction:

$$\frac{5}{8} = \frac{5 \times 3}{8 \times 3} = \frac{15}{24}$$

3. Check your answer by reducing it. You should arrive back at the original fraction.

$$\frac{15}{24} = \frac{15 \div 3}{24 \div 3} = \frac{5}{8}$$

Sample Question 2

A customer bought $\frac{2}{3}$ of a yard of fabric, but Katherine wants to convert this into 36ths so that she can see how many inches that is. Raise $\frac{2}{3}$ to 36ths.

Practice

Each of the following fractions is a measurement in lowest terms of how many yards of fabric various customers purchased. Help Katherine raise all these fractions into 36ths and then determine how many inches of fabric were purchased.

25. $\frac{5}{6}$

26. $\frac{1}{3}$

27. $\frac{7}{9}$

28. $\frac{5}{4}$

29. 3

Now raise each of these fractions to the higher terms that are indicated.

_____ **30.** $\frac{2}{9} = \frac{}{27}$

_____ **31.** $\frac{2}{5} = \frac{}{500}$

_____ **32.** $\frac{3}{10} = \frac{}{200}$

_____ **33.** $\frac{5}{6} = \frac{}{300}$

_____ **34.** $\frac{2}{9} = \frac{}{810}$

Comparing Fractions

Which fraction is larger, $\frac{3}{8}$ or $\frac{3}{5}$? Don't be fooled into thinking that $\frac{3}{8}$ is larger just because it has the larger denominator. There are several ways to compare two fractions, and they can be best explained by example.

- **Use your intuition: "pizza" fractions.** Let's return to Gina at the pizza place, who just took 2 medium cheese pizzas out of the oven. She cuts one of them into 5 slices and the second pizza into 8 slices. As you can see in the following illustration, the pizza that is divided into just 5 slices has *larger* slices. If you eat 3 of them, you're eating more pizza than if you eat 3 slices from the pizza with 8 slices. Thus, $\frac{3}{5}$ is larger than $\frac{3}{8}$.

  5 slices:

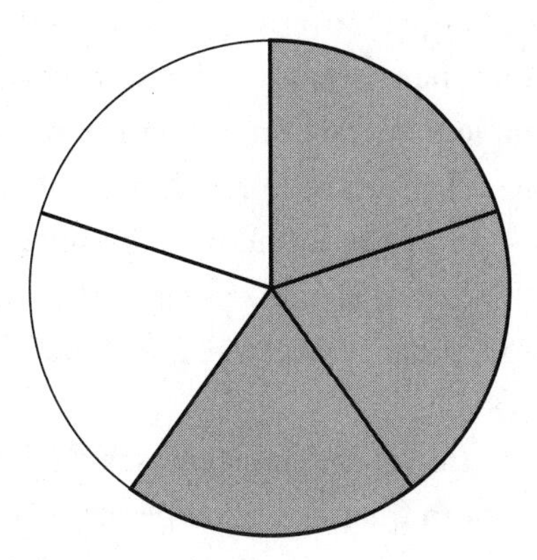

  8 slices:

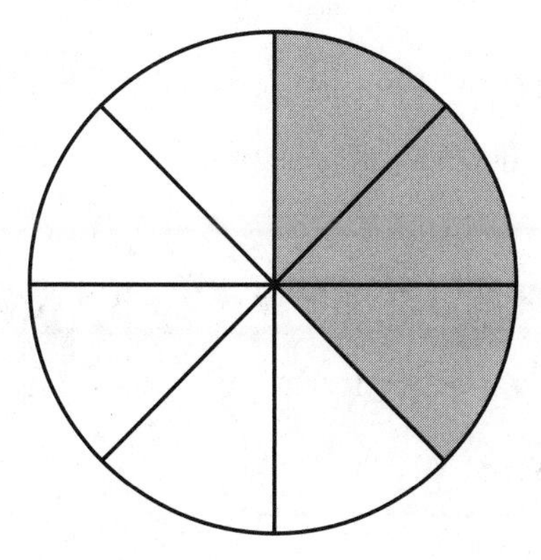

CAUTION

The "pizza" fraction method is not always the best method to rely on. This method is accurate when comparing fractions that have the same numerator, but different denominators. When the numerators are the same, the fraction with the larger denominator has been divided into smaller pieces, so that fraction represents the smaller number. When the denominators are different, choose one of the following methods.

- **Compare the fractions to $\frac{1}{2}$.** Both $\frac{3}{8}$ and $\frac{3}{5}$ are close to $\frac{1}{2}$. However, $\frac{3}{5}$ is more than $\frac{1}{2}$, while $\frac{3}{8}$ is less than $\frac{1}{2}$. Therefore, $\frac{3}{5}$ is larger than $\frac{3}{8}$. Comparing fractions to $\frac{1}{2}$ is actually quite simple. The fraction $\frac{3}{8}$ is less than $\frac{4}{8}$, which is the same as $\frac{1}{2}$; in a similar fashion, $\frac{3}{5}$ is more than $2\frac{1}{2}$, which is the same as $\frac{1}{2}$. ($2\frac{1}{2}$ may sound like a strange fraction, but you can easily see that it's the same as $\frac{1}{2}$ by considering a pizza cut into 5 slices. If you were to eat half the pizza, you'd eat $2\frac{1}{2}$ slices.)
- **Raise both fractions to higher terms with the same denominator.** When both fractions have the same denominator, then all you need to do is compare their numerators.

$$\frac{3}{5} = \frac{3 \times 8}{5 \times 8} = \frac{24}{40} \quad \text{and} \quad \frac{3}{8} = \frac{3 \times 5}{8 \times 5} = \frac{15}{40}$$

Since 24 is greater than 15, $\frac{24}{40}$ is greater than $\frac{15}{40}$.

This shows that $\frac{3}{5}$ is greater than $\frac{3}{8}$.

CAUTION

A fraction with the larger denominator isn't always smaller. For instance, $\frac{7}{8}$ is greater than $\frac{1}{5}$.

- **Shortcut: cross multiply.** "Cross multiply" means to perform diagonal multiplication between the denominators of each fraction with the numerators of the other fraction. To compare fractions through cross multiplication, write the cross multiplication results above the fractions and compare them. The cross product that was larger will be sitting above the larger fraction.

$$\overset{\textcircled{24}}{\frac{3}{5}} \;\times\; \overset{\textcircled{15}}{\frac{3}{8}} \quad \text{vs}$$

Since 24 is greater than 15, this indicates that $\frac{3}{5}$ is larger than $\frac{3}{8}$.

Practice

Peter needs to tell his boss which of the following fractions are the largest in each group. Help him out.

_____ **35.** $\frac{2}{5}$ *or* $\frac{3}{5}$

_____ **36.** $\frac{2}{3}$ *or* $\frac{4}{5}$

_____ **37.** $\frac{6}{7}$ *or* $\frac{7}{6}$

_____ **38.** $\frac{3}{10}$ *or* $\frac{3}{11}$

_____ **39.** $\frac{1}{5}$ *or* $\frac{1}{6}$

_____ **40.** $\frac{7}{9}$ *or* $\frac{4}{5}$

_____ **41.** $\frac{1}{3}$ *or* $\frac{2}{5}$ *or* $\frac{1}{2}$

_____ **42.** $\frac{5}{8}$ *or* $\frac{9}{17}$ *or* $\frac{18}{35}$

_____ **43.** $\frac{1}{10}$ *or* $\frac{10}{101}$ *or* $\frac{100}{1{,}001}$

_____ **44.** $\frac{3}{7}$ *or* $\frac{33}{77}$ *or* $\frac{9}{21}$

TRY THIS

It's time to take a look at your pocket change again! After you gather a pile of change, write the amount of change you have in the form of a fraction out of 100. Then reduce the fraction to its lowest terms.

You can do the same thing with time intervals that are less than an hour. How many minutes until you have to leave for work, go to lunch, or begin your next activity for the day? Express the time as a fraction out of 60 and then reduce it to lowest terms.

Answers

Practice Problems

1. 0.5
2. 0.25
3. 0.75
4. $0.\overline{3}$
5. $0.\overline{6}$
6. 0.125
7. 0.375
8. 0.625
9. 0.875
10. 0.2
11. 0.4
12. 0.6
13. 0.8
14. 0.1
15. $\frac{2}{9}$
16. $\frac{1}{6}$
17. $\frac{2}{1} = 2$
18. $\frac{2}{3}$
19. $\frac{15}{21}$
20. $\frac{4}{7}$
21. $\frac{1}{7}$
22. $\frac{1}{35}$
23. $\frac{1}{2}$
24. $\frac{1}{50}$
25. $\frac{30}{36}$
26. $\frac{12}{36}$
27. $\frac{28}{36}$

28. $\frac{45}{36}$
29. $\frac{108}{36}$
30. $\frac{6}{27}$
31. $\frac{200}{500}$
32. $\frac{60}{200}$
33. $\frac{250}{300}$
34. $\frac{180}{810}$
35. $\frac{3}{5}$
36. $\frac{4}{5}$
37. $\frac{7}{6}$
38. $\frac{3}{10}$
39. $\frac{1}{5}$
40. $\frac{4}{5}$
41. $\frac{1}{2}$
42. $\frac{5}{8}$
43. $\frac{1}{10}$
44. All equal

Sample Question 1

Divide the numerator and denominator by 6:

$$\frac{30 \div 6}{36 \div 6} = \frac{5}{6}$$

Sample Question 2

1. Divide the desired denominator (36) by the original denominator (3) to find the multiplication factor you will use to create the new fraction:

$$36 \div 3 = 12$$

2. Multiply the factor found in step 1 (12) by the numerator (2) and denominator (3) and write the products (multiplication answers) as a new fraction:

$$\frac{2}{3} = \frac{2 \times 12}{3 \times 12} = \frac{24}{36}$$

3. Check your answer by reducing it. You should arrive back at the original fraction.

$$\frac{24}{36} = \frac{24 \div 12}{36 \div 12} = \frac{2}{3}$$

Since $\frac{2}{3}$ is the same fraction as $\frac{24}{36}$, that means that $\frac{2}{3}$ of a yard is equivalent to 24 inches out of a 36-inch yard.

CHAPTER

3 Adding and Subtracting Fractions

I know that two and two make four—and should be glad to prove it too if I could—though I must say if by any sort of process I could convert two and two into five it would give me much greater pleasure.

—George Gordon, Lord Byron, British poet (1788–1824)

CHAPTER SUMMARY

In this chapter, you will learn how to add and subtract fractions and mixed numbers.

It is imperative for workers in all types of industries to be able to add and subtract fractions and mixed numbers. Carpentry, design, financial analysis, and stonemasonry are just some of the careers that require working with fractions on a daily basis. But is adding and subtracting fractions as simple as just adding or subtracting the numerators and denominators? For example, if Gina has $\frac{1}{2}$ of a cheese pizza and $\frac{1}{2}$ of a mushroom pizza, could we see how much pizza she has in total by just adding the two fractions straight across?

$$\frac{1}{2} + \frac{1}{2} \stackrel{?}{=} \frac{1+1}{2+2} \stackrel{?}{=} \frac{2}{4}$$

Since $\frac{2}{4}$ reduces to $\frac{1}{2}$, this method would determine that $\frac{1}{2} + \frac{1}{2} = \frac{1}{2}$ which makes no sense and is a false statement! Therefore we have just proven that we cannot simply add straight across to find the sum of two fractions. Now that we know how *not* to add fractions, let's take a look at how to *properly* add two fractions.

Adding Fractions

When adding fractions, you must make sure that the fractions have *common denominators* before moving forward. When two fractions have the same denominator, it means that all the pieces are the same size, and therefore you can just combine the numerators. When adding two fractions that have the same denominators, *add the numerators together and keep the denominator the same.* So, the initial problem we were considering, $\frac{1}{2} + \frac{1}{2}$,would be done as such:

$$\frac{1}{2} + \frac{1}{2} = \frac{1+1}{2} = \frac{2}{2}$$

And since $\frac{2}{2} = 1$, it follows that Gina has 1 full pizza remaining.

Now let's visualize the addition of fractions using the same "pizza slice" examples:

$\frac{1}{2}$ + $\frac{1}{2}$ = $\frac{2}{2} = 1$

Example

Eddie needs to fold over $\frac{2}{8}''$ of fabric and then make another fold of $\frac{4}{8}''$ for the hem of a costume he is making. How many inches of fabric in total will Eddie use to make the hem?

1. Since the fractions have common denominators, add the numerators and keep the denominator the same:

$$\frac{2}{8} + \frac{4}{8} = \frac{2+4}{8}$$

2. After adding, reduce the fraction to lowest terms if possible:

$$\frac{2+4}{8} = \frac{6}{8} = \frac{3}{4}$$

3. So Eddie will need $\frac{3}{4}''$ of fabric for the hem and the figure shows a visual illustration of this process:

$$\frac{2}{8} + \frac{4}{8} = \frac{6}{8} = \frac{3}{4}$$

CAUTION

Remember, do NOT add the denominators! Notice that $\frac{2}{5} + \frac{2}{5} \neq \frac{2+2}{5+5}$, because $\frac{2+2}{5+5} = \frac{4}{10}$, which reduces to $\frac{2}{5}$. It does not make sense that $\frac{2}{5} + \frac{2}{5}$ could equal $\frac{2}{5}$!

Sample Question 1

Eli needs to order granite to make a custom counter on either side of an outdoor grill. He needs one piece that is $\frac{5}{6}$ of a yard for the left side of the grill, and another piece that is $\frac{4}{6}$ of a yard for the right side of the grill. Draw a picture to illustrate $\frac{5}{6} + \frac{4}{6}$ and calculate how much granite Eli should order.

Finding the Least Common Denominator

When adding fractions with different denominators, it is necessary to raise some or all the fractions to higher terms so that they all have the same denominator. Once all the fractions have a **common denominator**, you will simply add the numerators and keep the denominator the same.

All the original denominators will divide evenly into the common denominator. If it is the smallest number into which they all divide evenly, it is called the **least common denominator (LCD)**. Addition is more efficient when the LCD is used than when any common denominator is used.

Here are some tips for finding the LCD:

- First check to see whether the smaller denominators can divide evenly into the largest denominator. If this is the case, you will just need to raise the smaller fractions to higher terms to have the LCD.
- If the first tip doesn't work, count by increasing multiples of the largest denominator until you find a denominator that all the other denominators can evenly divide. Then raise all the fractions to that denominator.

TIP

The fastest way to find a common denominator is to multiply the two denominators together. Example: For $\frac{1}{4}$ and $\frac{3}{8}$ you can use $4 \times 8 = 32$ as your common denominator.

Note: Notice that 32 is *not* the LCD. Although this method may appear to be the quickest, it will often not give you the LCD, which means you will need to do more work at the end of your problem to reduce your answer.

Example

$\frac{1}{6} + \frac{7}{12}$

1. Notice that the smaller denominator (6) evenly divides into the larger denominator (12), so find the LCD by raising $\frac{1}{6}$ to 12ths:

$$\frac{1}{6} = \frac{1 \times 2}{6 \times 2} = \frac{2}{12}$$

2. Add the numerators and keep the denominator the same:

$$\frac{2}{12} + \frac{7}{12} = \frac{9}{12}$$

3. Reduce to lowest terms or write as a mixed number, if applicable:

$$\frac{9}{12} = \frac{3}{4}$$

Example

$\frac{2}{3} + \frac{4}{5}$

1. Here 3 does not divide evenly into 5, so find a common denominator by multiplying the denominators by each other: $3 \times 5 = 15$
2. Raise each fraction to 15ths, the LCD: $\frac{2}{3} = \frac{10}{15}$, $\frac{4}{5} = \frac{12}{15}$
3. Add numerators and keep the denominator the same: $\frac{22}{15}$
4. Reduce to lowest terms or write as an equivalent mixed number, if applicable: $\frac{22}{15} = 1\frac{7}{15}$

Pictorially, this is represented by converting each fraction into an appropriately shaded pizza of 15 slices. Then add the numerators and keep the common denominator in your final answer.

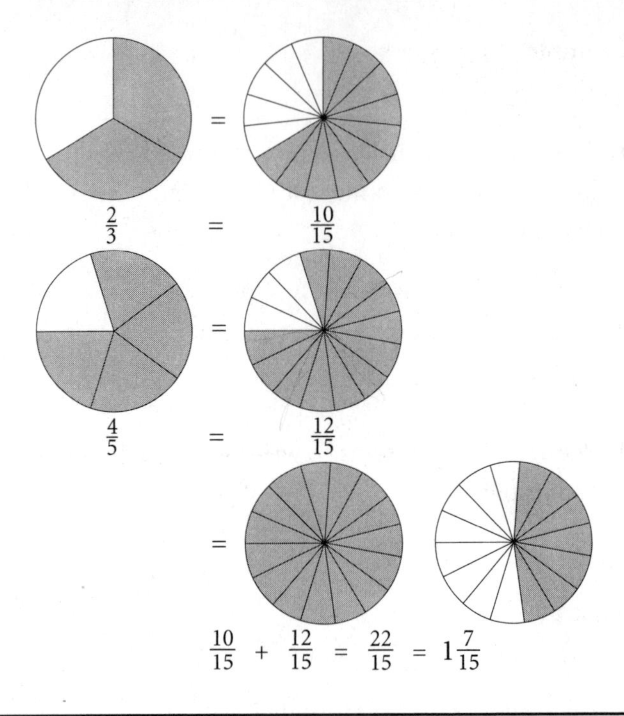

Sample Question 2

Gina sells $\frac{5}{8}$ of a mushroom pizza and $\frac{3}{4}$ of an onion pizza to the same family. How much pizza in all did the family buy? Draw a picture to represent the problem.

Adding Mixed Numbers

Mixed numbers, you remember, consist of a whole number and a proper fraction. To add mixed numbers:

1. Add the fractional parts of the mixed numbers. If the sum is an improper fraction, change it to a mixed number.
2. Add the whole number parts of the original mixed numbers.
3. Add the results of steps 1 and 2.

Example

$2\frac{3}{5} + 1\frac{4}{5}$

1. Add the fractional parts of the mixed numbers and change the improper fraction into a mixed number: $\frac{3}{5} + \frac{4}{5} = \frac{7}{5} = 1\frac{2}{5}$
2. Add the whole number parts of the original mixed numbers: $2 + 1 = 3$
3. Add the results of steps 1 and 2: $1\frac{2}{5} + 3 = 4\frac{2}{5}$

Sample Question 3

On Saturday Eli ordered $4\frac{2}{3}$ yards of granite and on Sunday he ordered another $1\frac{2}{3}$ yards. In total, how many yards of granite did Eli order over the weekend?

Practice

Add and reduce. Write your answer as either a proper fraction or a mixed number.

_____ **1.** $\frac{2}{5} + \frac{1}{5}$

_____ **2.** $\frac{3}{4} + \frac{1}{4}$

_____ **3.** $3\frac{1}{8} + 2\frac{3}{8}$

_____ **4.** $\frac{3}{10} + \frac{2}{5}$

_____ **5.** $3\frac{1}{2} + 5\frac{3}{4}$

_____ **6.** $2\frac{1}{3} + 3\frac{1}{2}$

_____ **7.** $\frac{5}{2} + 2\frac{1}{5}$

_____ **8.** $\frac{3}{10} + \frac{5}{8}$

_____ **9.** $1\frac{1}{5} + 2\frac{2}{3} + \frac{4}{15}$

_____ **10.** $2\frac{3}{4} + 3\frac{1}{6} + 4\frac{1}{12}$

Subtracting Fractions

As with addition, if the fractions you're subtracting have the same denominators, just subtract the numerators and keep the denominator the same.

Example

$\frac{4}{9} - \frac{3}{9} = \frac{4-3}{9} = \frac{1}{9}$

Visually, start with a pizza with 4 of 9 slices shaded. Then, erase 3 of those shaded slices. The fraction left is the difference. Pictorially, we have

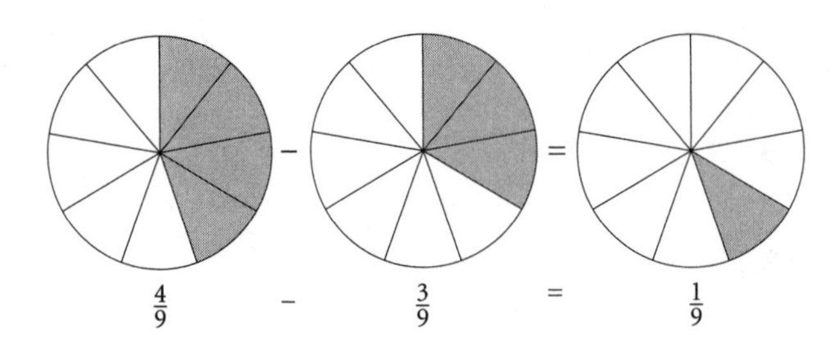

$\frac{4}{9} \quad - \quad \frac{3}{9} \quad = \quad \frac{1}{9}$

Sample Question 4

Eli goes to buy lunch for himself and his boss. He buys $\frac{5}{8}$ of a pizza and his boss eats $\frac{3}{8}$ of the pizza. Perform $\frac{5}{8} - \frac{3}{8}$ to see how much pizza is left for Eli. Draw a picture to represent your work.

Subtracting fractions with different denominators requires the same steps you learned for addition. First raise some or all the fractions to higher terms so that they all have the same denominator. Once all the fractions have a **common denominator**, you will simply subtract the numerators and keep the denominator the same. As with addition, subtraction is often more efficient if you use the LCD rather than a larger common denominator.

Example

$\frac{5}{6} - \frac{3}{4}$

1. Find the LCD. The easiest way to find the LCD is to check the multiplication table for 6, the larger of the two denominators. The smallest number into which both denominators divide evenly is 12.
2. Raise each fraction to 12ths, the LCD: $\frac{5}{6} = \frac{10}{12}$
3. Subtract as usual: $-\frac{3}{4} = \frac{9}{12}$

$$\frac{10}{12} - \frac{9}{12} = \frac{1}{12}$$

Sample Question 5

Eli had $\frac{3}{4}$ of a ton of gravel at the beginning of the week. His Monday project required $\frac{2}{5}$ of a ton of gravel. How much does he have remaining for Tuesday?

Subtracting Mixed Numbers

To subtract mixed numbers:

1. If the second fraction is smaller than the first fraction, subtract it from the first fraction. Otherwise, you'll have to "borrow" (explained by example further on) before subtracting fractions.
2. Subtract the second whole number from the first whole number.
3. Combine the results of steps 1 and 2 through addition.

Example

$4\frac{3}{5} - 1\frac{2}{5}$

1. Subtract the fractions: $\frac{3}{5} - \frac{2}{5} = \frac{1}{5}$
2. Subtract the whole numbers: $4 - 1 = 3$
3. Combine the results of steps 1 and 2: $\frac{1}{5} + 3 = 3\frac{1}{5}$

When the second fraction is bigger than the first fraction, you'll have to perform an extra "borrowing" step before subtracting the fractions, as illustrated in the following example.

Example

$7\frac{3}{5} - 2\frac{4}{5}$

1. You can't subtract the fractions in the present form because $\frac{4}{5}$ is bigger than $\frac{3}{5}$. So you have to "borrow":
   - Rewrite the 7 part of $7\frac{3}{5}$ as $6\frac{5}{5}$: (Note: Fifths are used because 5 is the denominator in $7\frac{3}{5}$; also, $6\frac{5}{5} = 6 + \frac{5}{5} = 7$.) $7 = 6\frac{5}{5}$
   - Then add back the $\frac{3}{5}$ part of $7\frac{3}{5}$: $7\frac{3}{5} = 7 + \frac{3}{5} = 6\frac{5}{5} + \frac{3}{5} = 6\frac{8}{5}$
2. Now you have a different, yet equivalent, version of the original problem: $6\frac{8}{5} - 2\frac{4}{5}$
3. Subtract the fractional parts of the two mixed numbers: $\frac{8}{5} - \frac{4}{5} = \frac{4}{5}$
4. Subtract the whole number parts of the two mixed numbers: $6 - 2 = 4$
5. Combine the results of the last 2 steps together: $4 + \frac{4}{5} = 4\frac{4}{5}$

TIP

Don't like the borrowing method previously shown? Here's another way to subtract mixed fractions:

- Change mixed fractions to improper fractions.
- Find common denominators.
- Subtract fractions: subtract the numerators and keep the denominator the same.
- If the answer is an improper fraction, change it back into a mixed number.

Sample Question 6

Katherine has $5\frac{1}{3}$ yards of red silk fabric left. Steve comes and buys $1\frac{3}{4}$ yards of it. How many yards of red silk fabric are remaining?

Practice

Subtract and reduce. Write your answer as either a proper fraction or a mixed number.

_____ **11.** $\frac{5}{6} - \frac{1}{6}$

_____ **12.** $\frac{7}{8} - \frac{3}{8}$

_____ **13.** $\frac{7}{15} - \frac{4}{15}$

_____ **14.** $\frac{2}{3} - \frac{3}{5}$

_____ **15.** $\frac{4}{3} - \frac{14}{15}$

_____ **16.** $\frac{7}{8} - \frac{1}{4} - \frac{1}{2}$

_____ **17.** $2\frac{4}{5} - 1$

_____ **18.** $3 - \frac{7}{9}$

_____ **19.** $2\frac{2}{3} - \frac{1}{4}$

_____ **20.** $2\frac{3}{8} - 1\frac{5}{6}$

TRY THIS

The next time you and a friend decide to pool your money together to purchase something, figure out what fraction of the whole each of you will contribute. Will the cost be split evenly: $\frac{1}{2}$ for your friend to pay and $\frac{1}{2}$ for you to pay? Or is your friend generous and offering to pay $\frac{2}{3}$ of the amount? If she pays $\frac{2}{3}$ of the cost of an item, what fraction of the cost will you pay? If you offer to pay $\frac{3}{4}$ of the cost, what fraction of the cost will she pay? Does the sum of the fractions add up to one? Can you afford to buy the item if your fractions don't add up to one?

Answers

Practice Problems

1. $\frac{3}{5}$
2. 1
3. $5\frac{1}{2}$
4. $\frac{7}{10}$
5. $9\frac{1}{4}$
6. $5\frac{5}{6}$
7. $4\frac{7}{10}$
8. $\frac{37}{40}$
9. $4\frac{2}{15}$
10. 10
11. $\frac{2}{3}$
12. $\frac{1}{2}$
13. $\frac{1}{5}$
14. $\frac{1}{15}$
15. $\frac{2}{5}$
16. $\frac{1}{8}$
17. $1\frac{4}{5}$
18. $2\frac{2}{9}$
19. $2\frac{5}{12}$
20. $\frac{13}{24}$

Sample Question 1

$$\frac{5}{6} + \frac{4}{6} = \frac{5+4}{6} = \frac{9}{6} = \frac{3}{2} = 1\frac{1}{2}$$

The result $\frac{9}{6}$ can be reduced to $\frac{3}{2}$, which can then be changed to the mixed number, $1\frac{1}{2}$. Here this is represented pictorially:

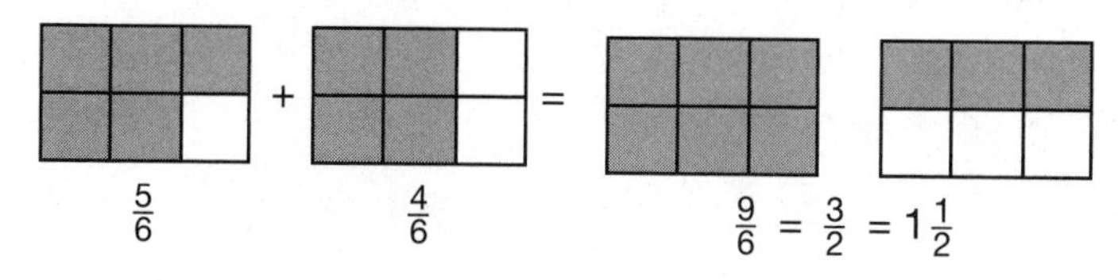

Sample Question 2

1. To find the LCD, check to see whether the smaller denominator (4) evenly divides the larger denominator (8). Since it does, you only need to raise the smaller fraction ($\frac{3}{4}$) to higher terms to have common denominators.
2. Raise $\frac{3}{4}$ to 8ths, the LCD: $\frac{3}{4} = \frac{6}{8}$
3. Add as usual: $\frac{5}{8} + \frac{6}{8} = \frac{11}{8}$
4. Change $\frac{11}{8}$ to a mixed number. $\frac{11}{8} = 1\frac{3}{8}$

Divide each piece of the pizza into two equal sub-slices, as follows:

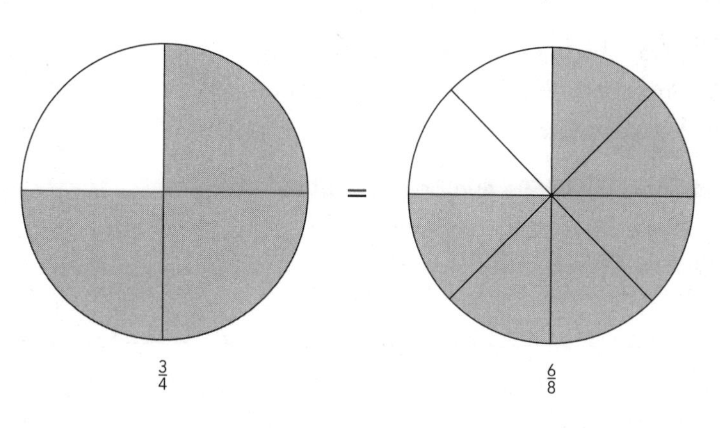

So, we add pictorially, as follows:

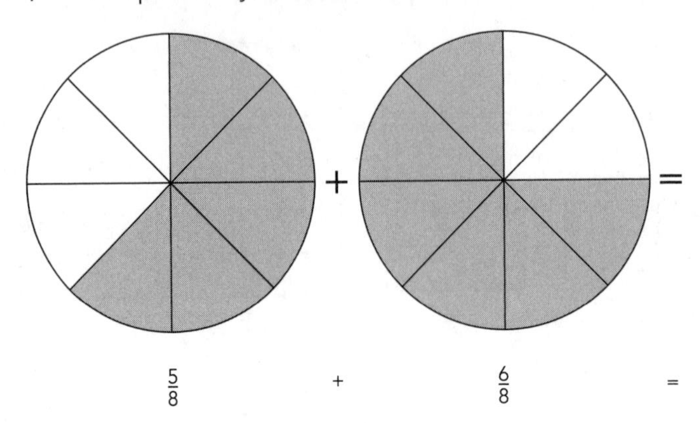

$\frac{5}{8}$ + $\frac{6}{8}$ =

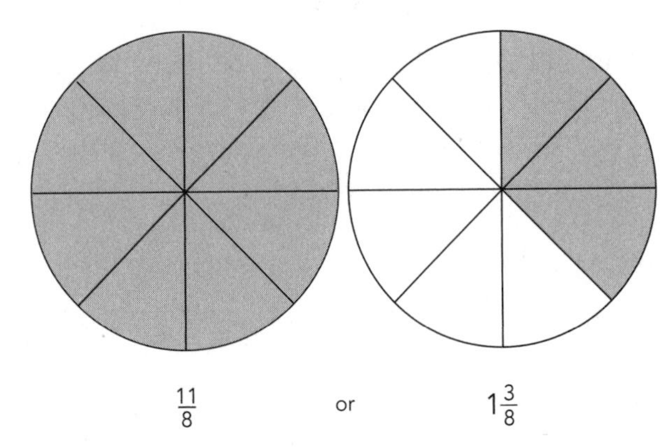

$\frac{11}{8}$ or $1\frac{3}{8}$

Sample Question 3

1. Add the fractional parts of the mixed numbers and change the improper fraction into a mixed number: $\frac{2}{3} + \frac{2}{3} = \frac{4}{3} = 1\frac{1}{3}$
2. Add the whole number parts of the original mixed numbers: $4 + 1 = 5$
3. Add the results of steps 1 and 2: $1\frac{1}{3} + 5 = 6\frac{1}{3}$

Sample Question 4

$\frac{5}{8} - \frac{3}{8} = \frac{5-3}{8} = \frac{2}{8}$, which reduces to $\frac{1}{4}$

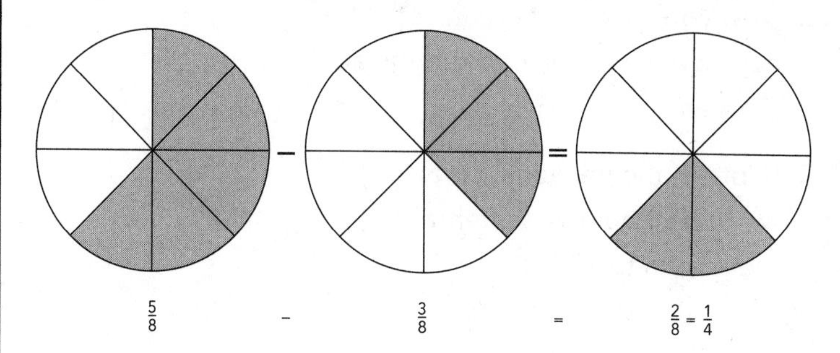

$\frac{5}{8}$ – $\frac{3}{8}$ = $\frac{2}{8} = \frac{1}{4}$

Sample Question 5

1. To find the LCD, check to see whether the smaller denominator (4) evenly divides the larger denominator (5). Since it does not, multiply the denominators together to find a common denominator: $4 \times 5 = 20$
2. Raise each fraction to 20ths, the LCD: $\frac{3}{4} = \frac{15}{20}$

   $-\frac{2}{5} = \frac{8}{20}$
3. Subtract as usual: $\frac{7}{20}$

Sample Question 6

1. You can't subtract the fractions in the present form because $\frac{3}{4}$ is bigger than $\frac{1}{3}$.

   So you have to "borrow":

   - Rewrite the 5 part of $5\frac{1}{3}$ as $4\frac{3}{3}$: $5 = 4\frac{3}{3}$

   (Note: Thirds are used because 3 is the bottom number in $5\frac{1}{3}$; also, $4\frac{3}{3} = 4 + \frac{3}{3} = 5$.)

- Then add back the $\frac{1}{3}$ part of $5\frac{1}{3}$: $5\frac{1}{3} = 5 + \frac{1}{3} = 4\frac{3}{3} + \frac{1}{3} = 4\frac{4}{3}$

2. Now you have a different, yet equivalent, version of the original problem: $4\frac{4}{3} - 1\frac{3}{4}$

3. Subtract the fractional parts of the two mixed numbers after raising them both to 12ths:

$$\begin{array}{r} \frac{4}{3} = \frac{16}{12} \\ -\frac{3}{4} = \frac{9}{12} \\ \hline \frac{7}{12} \end{array}$$

4. Subtract the whole number parts of the two mixed numbers: $4 - 1 = 3$

5. Add the results of the last two steps together: $3 + \frac{7}{12} = 3\frac{7}{12}$

CHAPTER

4 Multiplying and Dividing Fractions

If you ask your mother for one fried egg for breakfast and she gives you two fried eggs and you eat both of them, who is better in arithmetic, you or your mother?

—From "Arithmetic," by CARL SANDBURG, poet (1878–1967)

CHAPTER SUMMARY

This chapter focuses on multiplication and division with fractions and mixed numbers.

Many workplace environments require employees to be able to multiply and divide fractions on a daily basis. For instance, Mark works at a bakery with a famous scone recipe that calls for $4\frac{3}{4}$ cups of flour and $\frac{2}{3}$ cup of sugar. His boss wants him to cut the recipe in half on Wednesdays since that is their slowest day of the week. Remember Eli who works for the landscape architect? He has 2 tons of sand at the warehouse and his foreman wants to know how many sandboxes that can fill if each one needs $\frac{1}{4}$ ton of sand. And Katherine just cut 8 pieces of fabric that are each $3\frac{5}{8}$ yards long and she wants to quickly determine how much fabric in total her client just purchased.

The good news is that multiplying and dividing fractions is actually much easier than adding and subtracting them. For addition and subtraction you always have to find common denominators, but multiplication and division do NOT require common denominators. When you multiply fractions, you can simply multiply both the numerators and the denominators. To divide fractions, you invert the second fraction and multiply. Of course, there are extra steps when you get to multiplying and dividing mixed numbers.

Multiplying Fractions

Multiplication by a proper fraction is the same as finding a part of something. For instance, suppose Gina serves you a personal-sized pizza that is cut into 4 slices. Each slice represents $\frac{1}{4}$ of the pizza. If you eat $\frac{1}{2}$ of a slice, then you've eaten $\frac{1}{2}$ of $\frac{1}{4}$ of a pizza, which is the same as $\frac{1}{8}$ of the whole pizza.

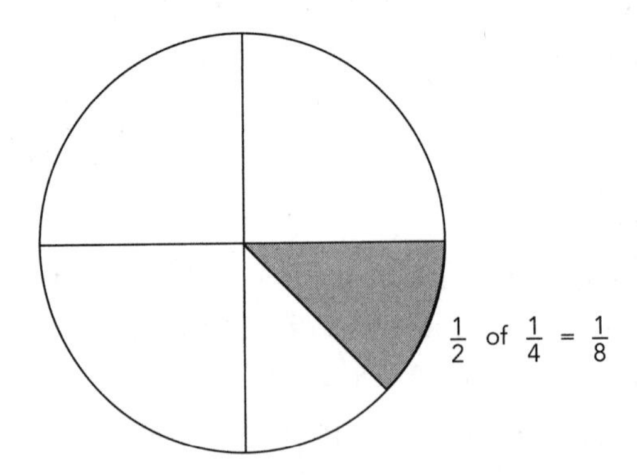

TIP

If you have 5 boxes *of* 6 cupcakes in each, how many cupcakes do you have? What if you have 4 boxes *of* 10 pens in each—how many pens do you have? What mathematical operation did you use to figure out the answers to these questions? It is helpful to realize that the word *of* translates to *multiplication* in math. That is why, when you ate $\frac{1}{2}$ *of* $\frac{1}{4}$ of a pizza, multiplication was used to calculate how much of the pizza you ate. This skill will be especially useful in word problems with fractions that might come up in your workplace.

Multiplying Fractions by Fractions

To multiply fractions:

1. Multiply the numerators together to get the numerator of the answer.
2. Multiply the denominators together to get the denominator of the answer.
3. Reduce and, if applicable, turn an improper fraction into a mixed number.

Example

$\frac{1}{2} \times \frac{1}{4}$

1. Multiply the numerators:
2. Multiply the denominators:

$$\frac{1 \times 1}{2 \times 4} = \frac{1}{8}$$

Example

$\frac{1}{3} \times \frac{3}{5} \times \frac{7}{4}$

1. Multiply the numerators:
2. Multiply the denominators:

$$\frac{1 \times 3 \times 7}{3 \times 5 \times 4} = \frac{21}{60}$$

3. Reduce: $\frac{21 \div 3}{60 \div 3} = \frac{7}{20}$

Now you try. Answers to sample questions are at the end of the lesson.

Sample Question 1

What is $\frac{2}{5}$ *of* $\frac{3}{4}$?

(Hint: If you are having trouble figuring out the math behind this, reread the TIP box before the example questions.)

Practice

Multiply and reduce.

_____ **1.** $\frac{1}{5} \times \frac{1}{3}$

_____ **2.** $\frac{2}{9} \times \frac{5}{4}$

_____ **3.** $\frac{7}{9} \times \frac{3}{5}$

_____ **4.** $\frac{3}{5} \times \frac{10}{7}$

_____ **5.** $\frac{3}{11} \times \frac{11}{12}$

_____ **6.** $\frac{4}{5} \times \frac{4}{5}$

_____ **7.** $\frac{2}{21} \times \frac{7}{2}$

_____ **8.** $\frac{9}{4} \times \frac{2}{15}$

_____ **9.** $\frac{5}{9} \times \frac{3}{13}$

_____ **10.** $\frac{8}{9} \times \frac{3}{12}$

Cancellation Shortcut

Sometimes you can cancel common factors before multiplying. Canceling is a shortcut that speeds up multiplication because you will not have to reduce the fraction to the lowest terms in the end. Canceling is similar to reducing: If there is a number that divides evenly into a numerator and a denomintor in any of the fractions being multiplied, do that division before multiplying. Up until now you have only reduced fractions by canceling vertically, however with multiplication you can cancel diagonally as well. Diagonal cancelation is what you will do most often when multiplying two fractions.

Example

$\frac{5}{6} \times \frac{9}{20}$

1. Cancel a 3 from both the 6 and the 9: $6 \div 3 = 2$ and $9 \div 3 = 3$. Cross out the 6 and the 9, as shown: $\frac{5}{\underset{2}{\cancel{6}}} \times \frac{\overset{3}{\cancel{9}}}{20}$
2. Cancel a 5 from both the 5 and the 20: $5 \div 5 = 1$ and $20 \div 5 = 4$. Cross out the 5 and the 20, as shown: $\frac{\overset{7}{\cancel{14}}}{3} \times \frac{11}{\underset{1}{\cancel{2}}}$
3. Multiply across the new numerators and the new denominators: $\frac{1 \times 3}{2 \times 4} = \frac{3}{8}$

When multiplying three or more fractions, the canceling shortcut can still be used between fractions that are not directly next to each other:

$$\frac{42}{9} \times \frac{4}{5} \times \frac{33}{21}$$

Nine and 4 cannot reduce and neither can 5 and 42. But you can reduce 9 and 33 by dividing by 3, and 42 and 21 can be reduced by dividing by 21:

$$\frac{\overset{2}{\cancel{42}}}{\underset{3}{\cancel{9}}} \times \frac{4}{5} \times \frac{\overset{11}{\cancel{33}}}{\underset{1}{\cancel{21}}} = \frac{(2 \times 4 \times 11)}{(3 \times 5 \times 1)} = \frac{88}{15}$$

Sample Question 2

Use cross cancellation to find what $\frac{4}{9}$ *of* $\frac{15}{22}$ is.

By the way, if you forget to cancel, don't worry. You'll still get the right answer, but you'll have to reduce it.

Practice

This time, cancel common factors diagonally before you multiply. If you do all the cancellations, you won't have to reduce your answer once you multiply.

_____ **11.** $\frac{1}{4} \times \frac{2}{3}$

_____ **12.** $\frac{2}{3} \times \frac{5}{8}$

_____ **13.** $\frac{8}{9} \times \frac{3}{16}$

_____ **14.** $\frac{21}{10} \times \frac{20}{63}$

_____ **15.** $\frac{300}{5{,}000} \times \frac{200}{7{,}000} \times \frac{100}{3}$

_____ **16.** $\frac{12}{36} \times \frac{27}{30}$

_____ **17.** $\frac{3}{7} \times \frac{14}{5} \times \frac{25}{6}$

_____ **18.** $\frac{2}{3} \times \frac{4}{7} \times \frac{3}{5}$

_____ **19.** $\frac{8}{13} \times \frac{52}{24} \times \frac{3}{4}$

_____ **20.** $\frac{1}{2} \times \frac{2}{3} \times \frac{3}{4} \times \frac{4}{5}$

_____ **21.** $\frac{5}{8} \times \frac{12}{30} \times \frac{4}{6} \times \frac{2}{3}$

_____ **22.** $\frac{20}{3} \times \frac{9}{100} \times \frac{5}{6} \times \frac{7}{2} \times \frac{1}{4}$

Multiplying Fractions by Whole Numbers

Before we jump into multiplying fractions by whole numbers, we need to remember that in order to represent a whole number in fraction format, we simply write the whole number over 1. Therefore, 4 is the same thing as $\frac{4}{1}$ and similarly, $8 = \frac{8}{1}$.

To multiply a fraction by a whole number:

1. Rewrite the whole number as a fraction with a denominator of 1.
2. Multiply as usual.

3. Reduce and, if applicable, turn an improper fraction into a mixed number.

Example

Katherine cut 5 pieces of fabric that were each $\frac{2}{3}$ of a yard. How much fabric did she cut? (Hint: What *is* 5 *of* $\frac{2}{3}$? Translate this problem to $5 \times \frac{2}{3}$ to solve.)

1. Rewrite 5 as a fraction: $5 = \frac{5}{1}$
2. Multiply the fractions: $\frac{5}{1} \times \frac{2}{3} = \frac{10}{3}$
3. Change the product $\frac{10}{3}$ to a mixed number. $\frac{10}{3} = 3\frac{1}{3}$

Sample Question 3

Mark made 24 scones and Mary bought $\frac{5}{8}$ of them. How many scones did Mary buy?

(Hint: What is $\frac{5}{8}$ *of* 24? Translate this word problem to $\frac{5}{8} \times 24$ to solve.)

Practice

Cancel common factors where possible, multiply, and then reduce. Convert products to mixed numbers where applicable.

_____ **23.** $12 \times \frac{3}{4}$

_____ **24.** $8 \times \frac{3}{10}$

_____ **25.** $3 \times \frac{5}{6}$

_____ **26.** $\frac{7}{24} \times 12$

_____ **27.** $16 \times \frac{7}{24}$

_____ **28.** $5 \times \frac{9}{10} \times 2$

_____ **29.** $60 \times \frac{1}{3} \times \frac{4}{5}$

_____ **30.** $\frac{1}{3} \times 24 \times \frac{5}{16}$

_____ **31.** $\frac{2}{5} \times 16 \times \frac{25}{32} \times 2$

_____ **32.** $\frac{1}{7} \times 5 \times \frac{2}{3} \times \frac{5}{3}$

Have you noticed that multiplying any number by a proper fraction produces an answer that's smaller than that number? It's the opposite of the result you get from multiplying whole numbers. That's because multiplying by a proper fraction is the same as finding a *part* of something.

Multiplying with Mixed Numbers

To multiply with mixed numbers, change each mixed number to an improper fraction and multiply.

Example

$4\frac{2}{3} \times 5\frac{1}{2}$

1. Change $4\frac{2}{3}$ to an improper fraction: $4\frac{2}{3} = \frac{4 \times 3 + 2}{3} = \frac{14}{3}$
2. Change $5\frac{1}{2}$ to an improper fraction: $5\frac{1}{2} = \frac{5 \times 2 + 1}{2} = \frac{11}{2}$
3. Multiply the fractions: Notice that you can cancel a 2 from both the 14 and the 2. $\frac{\overset{7}{\cancel{14}}}{3} \times \frac{11}{\underset{1}{\cancel{2}}}$
4. Change the improper fraction to a mixed number. $\frac{77}{3} = 25\frac{2}{3}$

Sample Question 4

Mark's scone recipe called for $1\frac{3}{4}$ cups of butter, but his boss wants him to just make $\frac{1}{2}$ of the recipe. How much butter should Mark use?

(Hint: How does $\frac{1}{2}$ *of* $1\frac{3}{4}$ translate into a math problem?)

CAUTION

When multiplying mixed numbers, is it a correct shortcut to multiply the whole numbers by each other and then multiply the fractions by each other? After all, that's what we did with addition and subtraction. Let's look at the following problem to see whether this could work: $5\frac{1}{10} \times 1\frac{9}{10}$. First, let's estimate what we think the product will be.

$\frac{1}{10}$ is a really small fraction, so let's round $5\frac{1}{10}$ to 5. $\frac{9}{10}$ is pretty close to a full whole, so let's round $1\frac{9}{10}$ to 2. We can now estimate that our answer should be close to $5 \times 2 = 10$. Now let's break up the multiplication and see whether our test answer is close to our estimate. If we multiply just the whole numbers by each other we get $5 \times 1 = 5$. When we multiply the fractions by each other we get $\frac{1}{10} \times \frac{9}{10} = \frac{9}{100}$. In this case, our combined answer would be $5\frac{9}{100}$. $5\frac{9}{100}$ is not at all close to our estimate of 10. Therefore, we see that it is NOT okay to break up the multiplication into parts; we must instead use improper fractions when multiplying mixed numbers.

Practice

Multiply and reduce. Change improper fractions to mixed or whole numbers.

_____ **33.** $2\frac{2}{3} \times \frac{2}{5}$

_____ **34.** $\frac{2}{11} \times 1\frac{3}{8}$

_____ **35.** $3 \times 2\frac{1}{3}$

_____ **36.** $1\frac{1}{5} \times 10$

_____ **37.** $\frac{5}{14} \times 4\frac{9}{10}$

_____ **38.** $5\frac{13}{5} \times \frac{5}{18}$

_____ **39.** $1\frac{1}{3} \times \frac{2}{3}$

_____ **40.** $8\frac{1}{3} \times 4\frac{4}{5}$

_____ **41.** $2\frac{1}{5} \times 4\frac{2}{3} \times 1\frac{1}{2}$

_____ **42.** $1\frac{1}{2} \times 2\frac{2}{3} \times 3\frac{3}{5}$

Dividing Fractions

When you divide a first number by a second number you are looking to see how many times the second number can fit into the first number. This is true of whether you are working with whole numbers or fractions. For instance, to find out how many $\frac{1}{4}$-pound pieces a 2-pound chunk of cheese can be cut into, you must divide

2 by $\frac{1}{4}$. As you can see from the following picture, a 2-pound chunk of cheese can be cut into eight $\frac{1}{4}$-pound pieces. ($2 \div \frac{1}{4} = 8$)

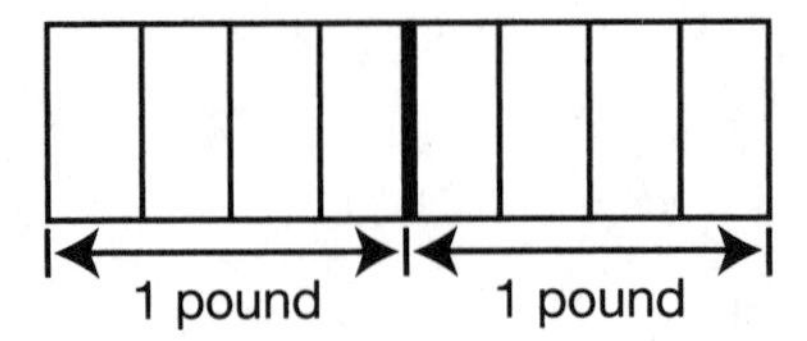

Reciprocal Fractions

The **reciprocal** of a fraction is the flipped version of that fraction, where the numerator and denominator have switched places. For example, $\frac{3}{5}$ is the reciprocal of $\frac{5}{3}$ and the reciprocal of 7 is $\frac{1}{7}$. It is important to be familiar with this term because reciprocals are used when dividing fractions.

Dividing Fractions by Fractions

To perform division between two numbers where at least one of them is a fraction, multiply the first number by the reciprocal of the second number. Therefore, a problem like $2 \div \frac{1}{4}$ will turn into $2 \times \frac{4}{1} = 8$. This isn't a magic rule that only applies to fractions—it also applies to whole numbers. Look at how $10 \div 2$ is the same thing as $10 \times \frac{1}{2}$: $10 \div 2 = 10 \times \frac{1}{2} = \frac{10}{1} \times \frac{1}{2} = \frac{10}{2} = 5$.

Although this rule *can* be used with whole numbers, it *must* be used with fractions and mixed numbers. Be careful that you are taking the reciprocal of the *second* number and not of the first number.

CAUTION

You are not allowed to cancel common factors in a division problem until it has been converted into a multiplication problem. After the reciprocal of the second number has been found and the division has been converted into multiplication, then you may diagonally cancel common factors.

Example

$\frac{1}{2} \div \frac{3}{5}$

1. Find the reciprocal of the second fraction ($\frac{3}{5}$): $\frac{5}{3}$
2. Change the ÷ into × and multiply the first number by the reciprocal of the second number: $\frac{1}{2} \times \frac{5}{3} = \frac{5}{6}$

Sample Question 5

$\frac{2}{5} \div \frac{3}{10}$

Another Format for Division

Sometimes fraction division is written in a different format. For example, $\frac{2}{1} \div \frac{3}{5}$ can also be written as $\frac{\frac{1}{2}}{\frac{3}{5}}$. This means the same thing! Regardless of the format used, the process of division and end result are the same.

TIP

Remember, when dividing a number by a positive fraction that is less than one, the answer is going to be larger than the original number. When dividing by an improper fraction (which has a value greater than one), your answer will be smaller than the original number. Use these facts to make sure your answers make sense.

Practice

Divide and reduce, canceling common factors where possible. Convert improper fractions to mixed or whole numbers.

_____ **43.** $\frac{4}{7} \div \frac{3}{5}$

_____ **44.** $\frac{2}{7} \div \frac{2}{5}$

_____ **45.** $\frac{1}{2} \div \frac{3}{4}$

_____ **46.** $\frac{5}{12} \div \frac{10}{3}$

_____ **47.** $\frac{1}{2} \div \frac{1}{3}$

_____ **48.** $\frac{5}{14} \div \frac{5}{14}$

_____ **49.** $\frac{9}{5} \div \frac{5}{9}$

_____ **50.** $\frac{45}{49} \div \frac{27}{35}$

_____ **51.** $\frac{35}{49} \div \frac{10}{21}$

_____ **52.** $\frac{7{,}500}{7{,}000} \div \frac{250}{140}$

Have you noticed that dividing a number by a proper fraction gives an answer that's larger than that number? It's the opposite of the result you get when dividing by a whole number.

Dividing Fractions by Whole Numbers or Vice Versa

To divide a fraction by a whole number or vice versa, change the whole number to a fraction by putting it over 1, and then divide as usual.

Example

$\frac{3}{5} \div 2$

1. Change the whole number (2) into a fraction: $2 = \frac{2}{1}$
2. Find the reciprocal of the second fraction ($\frac{2}{1}$): $\frac{1}{2}$
3. Change ÷ to × and multiply the two fractions: $\frac{3}{5} \times \frac{1}{2} = \frac{3}{10}$

For the next example we will use the same numbers, but switch the order of the division:

Example

$2 \div \frac{3}{5}$

1. Change the whole number (2) into a fraction: $2 = \frac{2}{1}$
2. Find the reciprocal of the second fraction ($\frac{3}{5}$): $\frac{5}{3}$
3. Change ÷ to × and multiply the two fractions: $\frac{2}{1} \times \frac{5}{3} = \frac{10}{3}$
4. Change the improper fraction to a mixed number: $\frac{10}{3} = 3\frac{1}{3}$

Did you notice that the *order* of division makes a difference? $\frac{3}{5} \div 2$ is not the same as $2 \div \frac{3}{5}$. But then, the same is true of division with whole numbers; $4 \div 2$ is not the same as $2 \div 4$.

⚠ CAUTION

When converting a division problem to a multiplication problem, remember to use the reciprocal of the second fraction and not the first.

Practice

Divide, canceling common factors where possible, and reduce. Change improper fractions into mixed or whole numbers.

_____ **53.** $2 \div \frac{3}{4}$

_____ **54.** $\frac{2}{7} \div 2$

_____ **55.** $1 \div \frac{3}{4}$

_____ **56.** $\frac{3}{4} \div 6$

_____ **57.** $\frac{8}{5} \div 4$

_____ **58.** $14 \div \frac{3}{14}$

_____ **59.** $\frac{25}{36} \div 5$

_____ **60.** $56 \div \frac{21}{11}$

_____ **61.** $35 \div \frac{7}{18}$

_____ **62.** $\frac{1,800}{12} \div 900$

_____ **63.** $\frac{1}{8} \div 8$

_____ **64.** $8 \div \frac{1}{8}$

_____ **65.** $\frac{\frac{2}{9}}{9}$

_____ **66.** $\frac{\frac{2}{9}}{2}$

Dividing with Mixed Numbers

With addition of mixed numbers, we learned how to add the whole numbers, then add the fractions, and lastly combine our answers to get a final sum. After some investigation, we found that when multiplying mixed numbers, this technique is NOT a correct shortcut. Instead, we must convert mixed numbers to improper fractions before multiplying. The same is true with division.

To divide with mixed numbers, change each mixed number to an improper fraction and then divide as usual.

Example

$2\frac{3}{4} \div \frac{1}{6}$

1. Change $2\frac{3}{4}$ to an improper fraction: $2\frac{3}{4} = \frac{2 \times 4 + 3}{4} = \frac{11}{4}$
2. Rewrite the division problem: $\frac{11}{4} \div \frac{1}{6}$
3. Find the reciprocal of the $\frac{1}{6}$ and multiply: $\frac{11}{\cancel{4}_{2}} \times \frac{\cancel{6}^{3}}{1} = \frac{11 \times 3}{2 \times 1} = \frac{33}{2}$
4. Change the improper fraction to a mixed number. $\frac{33}{2} = 16\frac{1}{2}$

⚠ CAUTION

Don't divide the whole parts and fraction parts separately when dividing mixed numbers.

Sample Question 6

One and a half tons of gravel must be divided between two construction worksites. Perform $1\frac{1}{2} \div 2$ to see how much gravel each worksite will get.

Practice

Divide, canceling common factors where possible, and reduce. Convert improper fractions to mixed or whole numbers.

_____ **67.** $2\frac{1}{2} \div \frac{3}{4}$

_____ **68.** $6\frac{2}{7} \div 11$

_____ **69.** $1 \div 1\frac{3}{4}$

_____ **70.** $2\frac{2}{3} \div \frac{5}{6}$

_____ **71.** $3\frac{1}{2} \div 3$

_____ **72.** $10 \div 4\frac{2}{3}$

_____ **73.** $1\frac{3}{4} \div 8\frac{3}{4}$

_____ **74.** $3\frac{2}{5} \div 6\frac{4}{5}$

_____ **75.** $2\frac{4}{5} \div 2\frac{1}{10}$

_____ **76.** $2\frac{3}{4} \div 1\frac{1}{2}$

_____ **77.** $\frac{5\frac{1}{3}}{10\frac{1}{3}}$

_____ **78.** $\frac{\frac{15}{2}}{3\frac{1}{2}}$

_____ **79.** $2\frac{5}{7} \div 2\frac{7}{5}$

_____ **80.** $1\frac{2}{3} \div 3\frac{1}{3}$

TRY THIS

Buy a small bag of candy. Before you eat any of the bag's contents, empty the bag and count how many pieces of candy are in it. Write down this number. Then walk around and collect three coworkers to share your candy. Now divide the candy equally among you. If the total number of candies you have is not divisible by four, you might have to cut some in half or quarters; this means you'll have to divide using fractions, which is great practice. Write down the mixed number of candy that each person received. The activity is an example of when dividing whole numbers can have a fractional outcome.

Answers

Practice Problems

1. $\frac{1}{15}$
2. $\frac{5}{18}$
3. $\frac{7}{15}$
4. $\frac{6}{7}$
5. $\frac{1}{4}$
6. $\frac{16}{25}$
7. $\frac{1}{3}$
8. $\frac{3}{10}$
9. $\frac{1}{9}$
10. $\frac{2}{9}$
11. $\frac{1}{6}$
12. $\frac{5}{12}$
13. $\frac{1}{6}$
14. $\frac{2}{3}$
15. $\frac{2}{35}$
16. $\frac{3}{10}$
17. 5
18. $\frac{8}{35}$
19. 1
20. $\frac{1}{5}$
21. $\frac{1}{9}$
22. $\frac{7}{16}$
23. 9
24. $2\frac{2}{5}$
25. $2\frac{1}{2}$
26. $3\frac{1}{2}$

27. $4\frac{2}{3}$

28. 9

29. 16

30. $2\frac{1}{2}$

31. 10

32. $\frac{50}{63}$

33. $1\frac{1}{15}$

34. $\frac{1}{4}$

35. 7

36. 12

37. $1\frac{3}{4}$

38. $2\frac{1}{9}$

39. $\frac{8}{9}$

40. 40

41. $15\frac{2}{5}$

42. $14\frac{2}{5}$

43. $\frac{20}{21}$

44. $\frac{5}{7}$

45. $\frac{2}{3}$

46. $\frac{1}{8}$

47. $1\frac{1}{2}$

48. 1

49. $\frac{81}{25}$

50. $1\frac{4}{21}$

51. $1\frac{3}{4}$

52. $\frac{3}{5}$

53. $2\frac{2}{3}$

54. $\frac{1}{7}$

55. $1\frac{1}{3}$
56. $\frac{1}{8}$
57. $\frac{2}{5}$
58. $65\frac{1}{3}$
59. $\frac{5}{36}$
60. $29\frac{1}{3}$
61. 90
62. $\frac{1}{6}$
63. $\frac{1}{64}$
64. 64
65. $\frac{2}{81}$
66. $\frac{1}{9}$
67. $3\frac{1}{3}$
68. $\frac{4}{7}$
69. $\frac{4}{7}$
70. $3\frac{1}{5}$
71. $1\frac{1}{6}$
72. $2\frac{1}{7}$
73. $\frac{1}{5}$
74. $\frac{1}{2}$
75. $1\frac{1}{3}$
76. $1\frac{5}{6}$
77. $\frac{16}{31}$
78. $2\frac{1}{7}$
79. $\frac{95}{119}$
80. $\frac{1}{2}$

Sample Question 1

Remember that *of* means multiplication, so $\frac{2}{5}$ *of* $\frac{3}{4}$ means $\frac{2}{5} \times \frac{3}{4}$.

1. Multiply the numerators: $2 \times 3 = 6$

2. Multiply the denominators: $5 \times 4 = 20$

3. Reduce: $\frac{6}{20} = \frac{3}{10}$

Sample Question 2

Remember that *of* means multiplication, so $\frac{4}{9}$ of $\frac{15}{22}$ means $\frac{4}{9} \times \frac{15}{22}$.

1. Cancel a 2 from both the 4 and the 22 $4 \div 2 = 2$ and $22 \div 2 = 11$. Cross out the 4 and the 22, as shown: $\frac{\overset{2}{\cancel{4}}}{9} \times \frac{15}{\underset{11}{\cancel{22}}}$

2. Cancel a 3 from both the 9 and the 15 $9 \div 3 = 3$ and $15 \div 3 = 5$. Cross out the 9 and the 15, as shown: $\frac{\overset{2}{\cancel{4}}}{\underset{3}{\cancel{9}}} \times \frac{\overset{5}{\cancel{15}}}{\underset{11}{\cancel{22}}}$

3. Multiply across the new numerators and the new denomintors: $\frac{2 \times 5}{3 \times 11} = \frac{10}{33}$

Sample Question 3

1. Rewrite 24 as a fraction: $24 = \frac{24}{1}$

2. Multiply the fractions:
Cancel an 8 from both the 8 and the 24. Then, multiply across the new numerators and the new denominators: $\frac{5}{\underset{1}{\cancel{8}}} \times \frac{\overset{3}{\cancel{24}}}{1} = \frac{15}{1} = 15$

Sample Question 4

1. Change $1\frac{3}{4}$ to an improper fraction: $1\frac{3}{4} = \frac{1 \times 4 \times 3}{4} = \frac{7}{4}$
2. Multiply the fractions: $\frac{1}{2} \times \frac{7}{4} = \frac{7}{8}$

Sample Question 5

1. Take the reciprocal of the second fraction ($\frac{3}{10}$): $\frac{10}{3}$
2. Change ÷ to × and multiply the first fraction by the new second fraction: $\frac{2}{\cancel{5}_1} \times \frac{\cancel{10}^2}{3} = \frac{4}{3}$
3. Change the improper fraction to a mixed number. $\frac{4}{3} = 1\frac{1}{3}$

Sample Question 6

1. Change $1\frac{1}{2}$ to an improper fraction: $1\frac{1}{2} = \frac{1 \times 2 + 1}{2} = \frac{3}{2}$
2. Change the whole number (2) into a fraction: $2 = \frac{2}{1}$
3. Rewrite the division problem: $\frac{3}{2} \div \frac{2}{1}$
4. Take the reciprocal of $\frac{2}{1}$ and multiply: $\frac{3}{2} \times \frac{1}{2} = \frac{3}{4}$

CHAPTER

5 Word Problems with Fractions

I have hardly ever known a mathematician who was capable of reasoning.

—PLATO, classical Greek philosopher (427 B.C.E.–347 B.C.E.)

CHAPTER SUMMARY

This final fraction chapter is devoted to word problems involving fractions.

This chapter reviews all the fraction lessons by presenting you with word problems. Fraction word problems are especially important because many situations in life do not involve only whole numbers. Think about it—how many of your friends are *exactly* 5 feet tall or 6 feet tall? How often does your fruit at the store weigh exactly 1 or 2 pounds? Most real-life scenarios involve fractions and decimals (which we will look at in the next lesson). Your boss is not likely to ask you what $1\frac{3}{4} \times \$50$ is, but you will need to be able to perform that multiplication if she asks you what the cost of $1\frac{3}{4}$ yards of \$50/yard fabric will be. Recognizing how to translate real-world situations involving fractions into math problems is the focus of this chapter.

Fractions with Measurement

Many applied fraction problems involve measurement of length, time, or weight. The key to success when working with measurement is to always consider how many smaller parts of a particular unit of measurement make a *whole.* Use that answer as your denominator when writing measurements as fractions. For example, when writing 8 inches as a fraction, since it takes 12 inches to make 1 foot, 8 inches is represented as $\frac{8}{12}$ of a foot.

Example

If there are 5,280 feet in a mile and Quinn walks 6,600 a day to his job, how many miles is Quinn's one-way walk to work?

1. Determine how many smaller parts make a whole (5,280 feet = 1 mile).
2. Use the answer from step 1 as the denominator in the fraction and the given information as the numerator: $\frac{6{,}600}{5{,}280}$
3. Reduce and, if applicable, turn the improper fraction into a mixed number, and remember to add the unit of measurement: $\frac{6{,}600}{5{,}280} = \frac{5}{4} = 1\frac{1}{4}$ miles.

Sample Question 1

Sasha ordered a bulk container of 480 ounces of mayonnaise to use in her food truck over the weekend. If a gallon is 128 ounces, express 480 ounces as gallons.

Practice

Convert the following measurements into the indicated fractional measurements.

_____ **1.** 50 ounces of sour cream expressed as cups (Note: 1 cup = 8 ounces)

_____ **2.** 85 cm expressed as meters (Note: 1 meter = 100 cm)

_____ **3.** 9,250 lbs. expressed as tons (Note: 1 ton = 2,000 lbs.)

_____ **4.** 10 hours expressed as days

_____ **5.** 40 minutes expressed as days Hint: How many minutes are in a day?)

Converting Fractional Measurements

To convert a fractional unit of measurement into a whole number measurement, multiply the fractional measurement by the number of smaller parts that make a whole. For example, if someone walks into a hardware store and orders $\frac{2}{3}$ of a foot of chain, the "of a foot" can be translated into "times 12" since there are 12 inches in a foot: $\frac{2}{3} \times 12 = \frac{2}{3} \times \frac{12}{1} = \frac{24}{3}$, which is 8 inches.

Example

Malia orders $\frac{5}{6}$ of a yard of decorative wrapping paper. How many inches has she ordered?

1. Determine how many smaller parts make a whole (36 inches = 1 yard).

2. Multiply the fractional measurement by your answer to step 1 and add the unit of measurement to your answer: $\frac{5}{6} \times 36 = 30$ inches.

Sample Question 2

A recipe calls for $\frac{7}{8}$ of a pound of shortening but Hailey wants to convert this into ounces for a cookbook she's writing. If there are 16 ounces in a pound, how many ounces of shortening does the recipe need?

1. Determine how many smaller parts make a whole (16 ounces = 1 pound).

2. Multiply the fractional measurement by your answer to step 1 and add the unit of measurement to your answer: $\frac{7}{8} \times 16 = 14$ ounces.

Practice

Convert the following fractional measurements into whole number measurements.

_____ **6.** How many feet are in $\frac{7}{8}$ of a mile?

_____ **7.** How many feet are in $2\frac{2}{3}$ yards?

_____ **8.** How many ounces are in $5\frac{1}{4}$ cups?

_____ **9.** How many minutes are in $\frac{5}{6}$ of an hour?

_____ **10.** How many inches are in $\frac{5}{8}$ of a foot?

CAUTION

When given mixed units that contain feet and inches, the number of inches is never added on as a decimal. 3 feet and 8 inches cannot be represented as 3.8 feet. Since 8 inches is $\frac{8}{12}$ of a foot, you can represent 3 feet and 8 inches as $3\frac{8}{12}$ feet. Or, you can choose to represent 3 feet and 8 inches in terms of inches. Convert the number of whole feet into inches by multiplying it by 12, and then add that to the 8 inches: 3 feet $= 3 \times 12 = 36$ inches, plus 8 inches, for 44 inches in total.

Word Problems

Now that we've reviewed how to represent measurements as fractions and how to convert fractional measurements into whole numbers, let's practice the arithmetic skills we learned in the first four chapters.

Each question group relates to one of the prior fraction chapters.

Writing and Comparing Fractions (Chapters 1 and 2)

_____ **11.** John worked 14 days in September. What fraction of the month did he work? Did he work more or less than Ella, who worked 26 out of the past 50 days?

_____ **12.** Clive's Cheesy Veggie dip calls for 3 ounces of cheese for the 15 ounce recipe and Clint's Creamy Veggie dip calls for 5 ounces of cheese in the 20 ounce recipe. Write both as fractions and determine which recipe is "cheesier."

_____ **13.** Alice lives 7 miles from her office. After driving 4 miles toward her office, Alice's car ran out of gas. What fraction of the trip had she already driven? What fraction of the trip remained?

_____ **14.** Mark had $10 in his wallet. He spent $6 for his lunch and left a $1 tip. What fraction of his money did he spend on his lunch, including the tip? Alisa spent $\frac{8}{11}$ of the money in her wallet on lunch. Who spent a greater part of his or her money on lunch?

_____ **15.** If Heather makes $2,000 a month and pays $750 for rent, what fraction of her income is spent on rent? Josie pays $1,300 of her $3,800 monthly paycheck on rent. Who spends a greater part of her income on rent?

_____ **16.** During a 30-day month, there were 8 weekend days and 1 paid holiday during which Marlene's office was closed. Marlene took off 3 days when she was sick and 2 days for personal matters. If she worked the rest of the days, what fraction of the month did Marlene work? Did she work more or less than $\frac{1}{2}$ the month?

Fraction Addition and Subtraction (Chapter 3)

_____ **17.** Stan drove $3\frac{1}{2}$ miles from home to work. He decided to go out for lunch and drove $1\frac{3}{4}$ miles each way to the local delicatessen. After work, he drove $\frac{1}{2}$ mile to stop at the cleaners and then drove $3\frac{2}{3}$ miles home. How many miles did he drive in total?

_____ **18.** An outside wall consists of $\frac{1}{2}$ inch of drywall, $3\frac{3}{4}$ inches of insulation, $\frac{5}{8}$ inch of wall sheathing, and 1 inch of siding. How thick is the entire wall, in inches?

_____ **19.** An invoice from Anawally Lumber indicates that your coworker bought $7\frac{2}{3}$ yards of border guard. How many feet of border guard will you have left over after using 18 feet?

_____ **20.** The length of a page in a particular book is 8 inches. The top and bottom margins are both $\frac{7}{8}$ inch. How long is the printed area *inside* the margins, in inches?

_____ **21.** The Boston Marathon is $26\frac{1}{5}$ miles long. At Heartbreak Hill, $20\frac{1}{2}$ miles into the race, how many miles remain?

_____ **22.** Howard bought 10,000 shares of VBI stock at $18\frac{1}{2}$ and sold it two weeks later at $21\frac{7}{8}$. How much did each share of stock increase in value? Express as a fraction.

_____ **23.** A window is 50 inches tall. To make curtains, Anya will need 2 more feet of fabric than the height of the window. How many yards of fabric will she need?

_____ **24.** Dr. Conlin discovers a tumor in her patient that is $4\frac{3}{10}$ micrometers in diameter. After 5 weeks of treatment, the tumor has shrunk to $2\frac{3}{4}$ micrometers. How many micrometers has the tumor shrunk? Dr. Conlin had hoped it would shrink at least $1\frac{1}{2}$ micrometers. Has she surpassed or fallen short of her treatment goal?

_____ **25.** Richard needs $12\frac{1}{2}$ pounds of fertilizer but has only $7\frac{5}{8}$ pounds. How many more pounds of fertilizer does he need?

Fraction Multiplication and Division (Chapter 4)

_____ **26.** A federal employee test is scored by adding 1 point for each correct answer and subtracting $\frac{1}{4}$ of a point for each incorrect answer. If Jan answered 31 questions correctly and 9 questions incorrectly, what was her score?

_____ **27.** A car's gas tank holds $10\frac{2}{5}$ gallons. How many gallons of gasoline are left in the tank when it is $\frac{1}{8}$ full?

_____ **28.** Four different work trucks need to evenly divide $6\frac{1}{2}$ tons of cement to take to the day's worksite. How many tons will each truck carry?

_____ **29.** How many $2\frac{1}{2}$-pound chunks of cheese can be cut from a single 20-pound piece of cheese?

_____ **30.** Each frame of a cartoon is shown for $\frac{1}{24}$ of a second. How many frames will be needed to make a cartoon that is $20\frac{1}{4}$ seconds long?

_____ **31.** A painting is $2\frac{1}{2}$ feet tall. To hang it properly, a wire must be attached exactly $\frac{1}{3}$ of the way down from the top. How many inches from the top should the wire be attached?

_____ **32.** A rope is cut in half and $\frac{1}{2}$ is discarded. From the remaining half, $\frac{1}{4}$ is cut off and discarded. What fraction of the original rope is left?

_____ **33.** Julio earns $14 an hour. When he works more than $7\frac{1}{2}$ hours a day, he gets overtime pay of $1\frac{1}{2}$ times his regular hourly wage for the extra hours. How much did he earn for working 10 hours in one day?

_____ **34.** Jodi earned $22.75 for working $3\frac{1}{2}$ hours. What was her hourly wage?

_____ **35.** A recipe for pizza dough calls for $3\frac{1}{2}$ cups of flour. How many cups of flour are needed to make only half the recipe?

_____ **36.** Of a journey, $\frac{3}{5}$ of the distance was covered on a plane and $\frac{1}{6}$ by driving. If the remainder of the trip was by boat, what fraction of the trip was by boat?

_____ **37.** Jane typed $1\frac{1}{2}$ pages of her paper in $\frac{1}{3}$ of an hour. At this rate, how many pages can she expect to type in 6 hours?

_____ **38.** Beto is barbecuing $\frac{1}{8}$-pound sliders as a Saturday evening special at his restaurant. If he anticipates that 30 of his clients will order the 2-slider appetizer and 40 of his clients will order the 3-slider entrée, how much meat should Beto buy for this special?

_____ **39.** Anita can make $3\frac{1}{2}$ necklaces per hour. If she goes into her office for $6\frac{1}{2}$ hours, pays bills for 45 minutes, takes a 30 minute lunch break, and makes necklaces for the rest of the time, how many necklaces will she make?

_____ **40.** Thy goes into a jewelry supply store and buys $4\frac{1}{2}$ yards of silver chain that costs $5.50 per *foot*. How much will he pay for this chain?

TRY THIS

Applied fraction problems are all around—you just need to look for them! The next time you buy fruit at the supermarket by the pound, see how many $\frac{1}{4}$–pound servings you could divide your entire purchase into. This will show you how many servings of fruit you have. Then, when you buy meat or fish, see how many $\frac{1}{3}$–pound servings you can divide your entire purchase into. This will show you how many entree-sized servings of protein you have.

Answers

Practice Problems

1. $6\frac{1}{4}$ cups
2. $\frac{17}{20}$ meter
3. $4\frac{5}{8}$ tons
4. $\frac{5}{12}$ day
5. $\frac{1}{36}$ day
6. 4,620 feet
7. 8 feet
8. 42 ounces
9. 50 minutes
10. $7\frac{1}{2}$ inches
11. John = $\frac{7}{15}$, Ella = $\frac{13}{25}$. Ella worked more.
12. Clive's Cheesy Veggie dip = $\frac{1}{5}$ cheese, Clint's Creamy Veggie dip = $\frac{1}{4}$. Clint's is "cheesier."
13. Driven = $\frac{4}{7}$, remained = $\frac{3}{7}$
14. Mark spent $\frac{7}{10}$, Alisa spent $\frac{8}{11}$. Alisa spent more of her money.
15. Heather = $\frac{15}{44}$, Josie = $\frac{13}{38}$. Heather spends a greater part of her paycheck on rent.
16. Marlene worked $\frac{8}{15}$, which is more than half the month.
17. $11\frac{1}{6}$ miles
18. $5\frac{7}{8}$ inches
19. 5 feet
20. $6\frac{1}{4}$ inches
21. $5\frac{7}{10}$ miles
22. $3\frac{3}{8}$
23. $2\frac{1}{18}$ yards
24. $1\frac{11}{20}$, surpassed

25. $4\frac{7}{8}$ pounds

26. $28\frac{3}{4}$

27. $1\frac{3}{10}$ gallons

28. $1\frac{5}{8}$ tons

29. 8 chunks

30. 486 frames

31. 10 inches

32. $\frac{3}{8}$

33. $157.50

34. $6.50/hour

35. $1\frac{3}{4}$ cups

36. $\frac{7}{30}$

37. 27 pages

38. $22\frac{1}{2}$ pounds

39. $18\frac{3}{8}$ necklaces

40. $74.25

Sample Question 1

1. Determine how many smaller parts make a whole (128 ounces = 1 gallon).
2. Use the answer from step 1 as the denominator in the fraction: $\frac{480}{128}$
3. Reduce and, if applicable, turn the improper fraction into a mixed number, and remember to add the unit of measurement: $\frac{480}{128} = \frac{15}{4} = 3\frac{3}{4}$ gallons.

Sample Question 2

1. Determine how many smaller parts make a whole (16 ounces = 1 pound).
2. Multiply the fractional measurement by your answer to step 1 and add the unit of measurement to your answer: $\frac{7}{8} \times 16 = 14$ ounces.

CHAPTER

6 Introduction to Decimals

The highest form of pure thought is in mathematics.

—PLATO, classical Greek philosopher (427 B.C.E.–347 B.C.E.)

CHAPTER SUMMARY

This first decimal chapter is an introduction to the concept of decimals. It explains the relationship between decimals and fractions, teaches you how to compare decimals, and gives you a tool called rounding for estimating decimals.

A decimal is another way to represent a fraction. You use decimals every day when you deal with measurements or money. For instance, $10.35 is a decimal that represents 10 dollars and 35 cents. The decimal point separates the whole dollars from the cents, which are fractions of dollars. Because there are 100 cents in one dollar, 1¢ is $\frac{1}{100}$ of a dollar, or $0.01; 10¢ is $\frac{10}{100}$ of a dollar, or $0.10; 25¢ is $\frac{25}{100}$ of a dollar, or $0.25; and so forth. In terms of measurements, a weather report might indicate that 2.7 inches of rain fell in 4 hours, you might drive 5.8 miles to the intersection of the highway, or the population of the United States might be estimated to grow to 374.3 million people by a certain year.

If there are non-zero numbers located on both sides of the decimal point, like 6.17, the number is called a **mixed decimal**. The numbers to the left of the decimal point represent whole numbers and the numbers to the right of the decimal represent a fraction of a number less than 1. If there are no whole numbers and only digits to the right of the decimal, like 0.17, this is called a **decimal** and its value is always less than 1. A whole number, like 6, is understood to have a decimal point to the right of the number and it is not necessary to write the decimal point. On the other hand, the decimal point *always* has to be present when representing a number less than 1, like 0.17, and although sometimes the zero is omitted and the number may be written as .17, it is preferable to leave the 0 to the left of the decimal point for clarity.

Decimal Names

Each decimal digit to the right of the decimal point has a special name. Here are the first four:

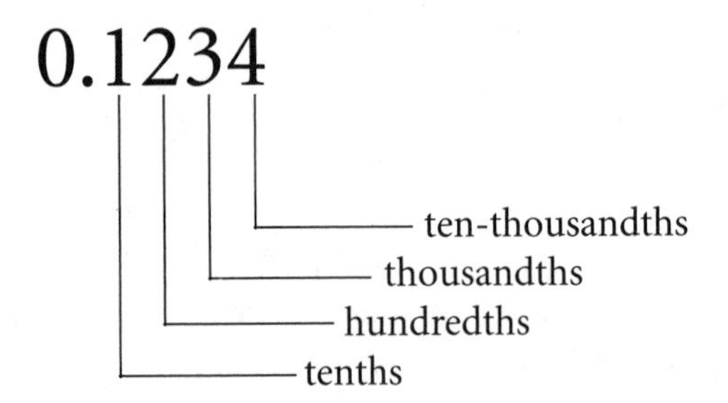

The digits have these names for a very special reason: The names reflect their fraction equivalents.

$0.1 = 1 \text{ tenth} = \frac{1}{10}$
$0.02 = 2 \text{ hundredths} = \frac{2}{100}$

$0.003 = 3$ thousandths $= \frac{3}{1,000}$
$0.0004 = 4$ ten-thousandths $= \frac{4}{10,000}$

As you can see, decimal names are ordered by multiples of 10: 10ths, 100ths, 1,000ths, 10,000ths, 100,000ths, 1,000,000ths, and so on. Be careful not to confuse decimal names with whole number names, which are very similar (tens, hundreds, thousands, etc.). The naming difference can be seen in the ths, which are used only for decimal digits.

Reading a Decimal

Here's how to read a mixed decimal, for example, 6.017:

1. The number to the left of the decimal point is a whole number. Just read that number as you normally would:	6
2. Say the word "and" for the decimal point:	and
3. The number to the right of the decimal point is the decimal value. Just read it:	17
4. The number of places to the right of the decimal point tells you the decimal's name. In this case, there are three places:	thousandths

Thus, 6.017 is read as *six and seventeen thousandths*, and its fraction equivalent is $6\frac{17}{1,000}$.

Here's how to read a decimal, for example, 0.28:

1. Read the number to the right of the decimal point:	28
2. The number of places to the right of the decimal point tells you the decimal's name. In this case, there are two places:	hundredths

Thus, 0.28 (or .28) is read as *twenty-eight hundredths*, and its fraction equivalent is $\frac{28}{100}$.

Informally, you could also read 0.28 as *point two eight*, but it doesn't quite have the same intellectual impact as *28 hundredths!*

Adding Zeros

Adding zeros to the very end of the decimal portion of a mixed decimal does not change its value. For example, 6.17 has the same value as each of these other mixed decimals:

6.17 =
6.170 =
6.1700 =
6.17000, and so forth

Remembering that a whole number is assumed to have a decimal point at its right, the whole number 6 has the same value as each of these:

6.
6.0
6.00
6.000, and so forth

> **⚠ CAUTION**
>
> Inserting any zeros to the right of the decimal point, but *before* the very end of the decimal portion, does change the value of a number. For example, 6.17 does not have the same value as any of these other mixed decimals:
>
> $$6.17 \neq$$
> $$6.017 \neq$$
> $$6.107 \neq$$
> $$6.1007 \neq$$
> $$60.17\text{, and so forth}$$

Practice

You are typing up your boss's research paper and he is submitting it to a journal that requires all measurements to be expressed in words, not numbers. Write each of the following in words:

1. 0.1 ______________________________

2. 0.01 ______________________________

3. 0.001 ______________________________

4. 0.0001 ______________________________

5. 0.00001 ______________________________

6. 5.19 ______________________________

7. 1.0521 ________________________________

8. 10.0000010 ________________________________

Your lab received these measurements over the phone from a scientist in Beijing, and your boss wants these measurements in number format. Write the following as decimals or mixed decimals.

_____ **9.** Six tenths

_____ **10.** Six hundredths

_____ **11.** Twenty-five thousandths

_____ **12.** Three hundred twenty-one thousandths

_____ **13.** Nine and six thousandths

_____ **14.** Three and one ten-thousandth

_____ **15.** Fifteen and two hundred sixteen thousandths

_____ **16.** One and one hundred one ten millionths

TIP

Decimals are all around us! They are used in money, measurement, and time, so it's important to read this section carefully and make sure you feel comfortable with them. Using decimals is essential in mastering practical, workplace, and real-world math skills.

Changing Decimals and Mixed Decimals to Fractions

To change a decimal to a fraction:

1. Write the digits of the decimal as the top number of a fraction.
2. Write the decimal's name as the bottom number of the fraction.

Example

Change 0.018 to a fraction.

1. Write 18 as the top of the fraction: $\frac{18}{}$
2. Since there are three places to the right of the decimal, it's thousandths.
3. Write 1,000 as the bottom number: $\frac{18}{1{,}000}$
4. Reduce by dividing 2 into the top and bottom numbers: $\frac{18 \div 2}{1{,}000 \div 2} = \frac{9}{500}$

Now try this sample question. Step-by-step solutions to sample questions are at the end of the lesson.

Sample Question 1

Change the mixed decimal 2.7 to a mixed number.

Practice

You work in a lab in New York that is occasionally invited to collaborate with a prestigious lab in Chicago. Your lab records weights in decimal format, but the Chicago lab has asked to see your findings as mixed fractions in lowest terms. Prepare all these numbers for the lab in Chicago.

_____ **17.** 0.1

_____ **18.** 0.03

_____ **19.** 0.75

_____ **20.** 0.99

_____ **21.** 0.005

_____ **22.** 0.125

_____ **23.** 0.046

_____ **24.** 5.04

_____ **25.** 4.15

_____ **26.** 123.45

_____ **27.** 20.0050

_____ **28.** 10.10005

Changing Fractions to Decimals

The bar that separates the numerator from the denominator in fractions can actually be read as "divided by." So $\frac{10}{5}$ can be read as "10 divided by 5," and $\frac{1}{4}$ is the same thing as "1 divided by 4." Therefore, fractions are turned into decimals by using long division. The most important thing to remember is that since the numerator is getting divided by the denominator, the numerator goes into the long division box and the denominator stays on the outside. To change a fraction to a decimal:

1. Set up a long division problem where the numerator is the *dividend* in the division box and the denominator is the *divisor* on the outside—but don't divide just yet.
2. Put a decimal point to the right of the dividend in the division box, followed by a few zeros.
3. Bring the decimal point straight up into the area for the answer (the *quotient*).
4. Divide.

Example

Change $\frac{3}{4}$ to a decimal.

1. Set up the division problem: $4\overline{)3}$
2. Add a decimal point and zeros to the divisor (3): $4\overline{)3.00}$
3. Bring the decimal point up into the answer: $4\overline{)3\overset{.}{\uparrow}00}$
4. Divide:

$$\begin{array}{r} 0.75 \\ 4\overline{)3.00} \\ \underline{28} \\ 20 \\ \underline{20} \\ 0 \end{array}$$

Thus, $\frac{3}{4}$ = 0.75, or 75 hundredths.

The same approach works when converting mixed numbers and improper fractions into decimals. In the case of mixed numbers, hold off on the whole number part and tack it onto the left side of the decimal point once you have performed the division.

For an improper fraction, the same approach works, but expect there to be nonzero digits *before* the decimal point.

Example
Change $\frac{26}{5}$ to a decimal.

1. Set up the division problem: $5\overline{)26}$
2. Add a decimal point and zeros to the dividend: $5\overline{)26.00}$
3. Bring the decimal point up into the answer: $5\overline{)26.00}$
4. Divide:

$$\begin{array}{r} 5.2 \\ 5\overline{)26.00} \\ \underline{-25} \\ 10 \\ \underline{-10} \\ 0 \end{array}$$

Note: In this example only one of the zeros was needed to complete the division, but in other examples 3 or 4 zeros might be needed.

Sample Question 2

Change $\frac{1}{5}$ to a decimal.

Repeating Decimals

Some fractions may require you to add more than two or three decimal zeros in order for the division to come out evenly. In fact, when you change a fraction like $\frac{2}{3}$ to a decimal, you'll keep adding decimal zeros until you're blue in the face because the division will never come out evenly! As you divide 3 into 2, you'll keep getting 6's:

$$\begin{array}{r} 0.6666\,etc. \\ 3\overline{)2.0000etc.} \\ \underline{18} \\ 20 \\ \underline{18} \\ 20 \\ \underline{18} \\ 20 \\ \underline{18} \\ 20 \end{array}$$

A fraction like $\frac{2}{3}$ becomes a *repeating decimal.* Its decimal value can be written as $0.\overline{6}$, or it can be approximated as 0.66, 0.666, 0.6666, and so forth. Its value can also be approximated by *rounding* it to 0.67 or 0.667 or 0.6667, and so forth. (Rounding is covered later in this lesson.)

If you have *fractionphobia* and panic when you have to do fraction arithmetic, it is fine to convert each fraction to a decimal and do the arithmetic in decimals. Warning: This should be a means of last resort—fractions are so much a part of daily living that it's important to be able to work with them.

Practice

The division bar on your keyboard is broken, but you have to enter all these measurements into the computer before you can leave work. Change them all to decimals.

_____ **29.** $\frac{2}{5}$

_____ **30.** $\frac{7}{10}$

_____ **31.** $\frac{1}{6}$

_____ **32.** $\frac{5}{7}$

_____ **33.** $\frac{4}{9}$

_____ **34.** $4\frac{3}{4}$

_____ **35.** $3\frac{2}{7}$

_____ **36.** $\frac{65}{4}$

_____ **37.** $\frac{107}{6}$

_____ **38.** $\frac{101}{3}$

Comparing Decimals

Decimals are easy to compare when they have the same number of digits after the decimal point. Just cover up the decimal point and compare the numbers to each other: For instance, you can compare 0.20 and 0.09 by comparing 20 to 9. Obviously 20 is bigger. When two decimals have a different number of digits to the right of the decimal, add zeros to the *end* of the shorter decimal until they have the same number of digits. (Recall that adding zeros in this manner does not change the value of a decimal.) Once both decimals have equal digits, ignore the decimal points and compare the numbers as we did with 0.20 and 0.09.

Example

Compare 0.08 and 0.1.

1. Since 0.08 has two decimal digits, tack one zero onto the end of 0.1, making it 0.10
2. To compare 0.10 to 0.08, just compare 10 to 8. Ten is larger than 8, so 0.1 is larger than 0.08.

CAUTION

Don't be tempted into thinking 0.08 is larger than 0.1 just because the whole number 8 is larger than the whole number 1. Remember that $0.08 = \frac{8}{100}$, while $0.1 = 0.10$, which is $\frac{10}{100}$. Since 10 hundredths is larger than 8 hundredths, 0.1 is larger than 0.08.

Sample Question 3

Put these decimals in order from least to greatest: 0.1, 0.11, 0.101, and 0.0111.

Practice

Order each group from lowest to highest.

_____ **39.** 0.2, 0.05, 0.009

_____ **40.** 0.417, 0.422, 0.396

_____ **41.** 0.019, 2.009, 0.01

_____ **42.** 0.82, 0.28, 0.8, 0.2

_____ **43.** 0.3, 0.30, 0.300

_____ **44.** 0.5, 0.05, 0.005, 0.505

_____ **45.** 1.1, 10.001, 1.101, 1.1000001

_____ **46.** 2.3, 0.230, 0.02–3–, 0.2–3–

_____ **47.** $\frac{1}{10}$, 0.01, $0.\overline{1}$, $\frac{11}{100}$, 0.011

_____ **48.** $\frac{1}{2}$, 0.05, $\frac{55}{100}$, 0.49, $\frac{480}{1,000}$

Rounding Decimals

Rounding a decimal is a way of *estimating* its value using fewer digits. You probably round numbers all the time, without even realizing it. If Gina sold you a slice of pizza for $3.89, you might tell your friend that your pizza cost around $4. If your manager asks what the sales were in the past hour, and the exact number is $389.40, you might tell her "around $400." Rounding is useful for making quick calculations when the exact answer is not necessary. A printed fabric costs $4.85/yard and Abe wants to purchase 11.25 yards of it, but his boss gave him a spending limit of $100. He could quickly estimate the cost like this:

> $4.85 is closer to $5 than it is to $4
> 11.25 yards is closer to 11 than it is to 12

Therefore, an estimated cost is $5 × 11 = $55, so Gabe will be under his $100 spending limit.

Since in this case, \$4.85 went *up* to \$5, we say it was **rounded up**. Since 11.25 went *down* to 11, we say it was **rounded down.**

Rounding is also used to simplify single figures when the exact number is not critical. If the CEO of the company you work for earns a salary of \$388,742, you might exclaim to a coworker that you can't believe the CEO makes almost \$400,000. If the population of a country is 27,240,892, it is closer to 27-million than it is to 28 million, so it might be cited in an article as a country "with an approximate population of 27 million." (In this case, it would not be extremely accurate to round 27 million up to 30 million since that would be adding an additional 3 million people (which is a lot), so it is more reasonable to keep it at 27 million.)

Rounding is an excellent way to see how *reasonable* your answer is to an arithmetic problem involving decimals. Let's say that when Gage was purchasing 11.25 yards of fabric at \$4.85/yard, the sales associate quickly entered it into the cash register and told him the total was \$545.63. Rather than canceling his purchase and telling the sales associate that he couldn't afford this fabric, Gage could compare \$545.63 to his estimate of \$55 and see that the associate must have made an error with the decimal point! When working with decimals, it's great to get in the habit of making an estimate to an arithmetic problem first, so that you can compare this estimate to the actual answer. If they are not at least reasonably similar, check your arithmetic.

To Round Up or Down—That Is the Question!

You already learned that numbers can be *rounded up* or *rounded down.* Numbers can also be rounded to any place value, depending on the size of the number and the real-world context to which it applies. Think of rounding as "cleaning up" a number while keeping its general value the same. When rounding small numbers, it is important not to make huge changes that drastically alter the number, but when dealing with larger numbers, the smaller digits are less significant. For example, if James was going to pay you

$12,042 for a used car, it would be reasonable for the person selling it to ignore the $42 and accept $12,000 for the car. However, if Kerry was going to pay you $42 for a used bike, it would be silly for the person selling it to ignore the $42 and accept $0 for the bike. Furthermore, it would be odd to round the $12,042 to $12,040, but it wouldn't be strange to round the $42 bike to $40. Rounding numbers often involves this type of common sense.

Rules of Rounding

On the job you will often need to round a number to the nearest **whole number.**

This means that the number will have no digits to the right of the decimal. Other times it may be required that you round to a specific place value, like to the nearest tenth or to the nearest hundredth. There will also be cases on the job when you will need to round to the place value that makes the most sense according to the context. Here are the two general rules for rounding numbers:

1. If there is no rounding place value specified, first decide which is the last place value you will keep.
2. Investigate the next digit *to the right* of the last place value that will be kept.

Case 1: If the digit to the right of your last desired place value is **4 or less**, round down by keeping your last desired digit the same and replacing all the digits to the right of it with zeros. (If you are rounding to a place value to the right of the decimal point, you can just drop the remaining digits to the right of it.)

Case 2: If the digit to the right of your last desired place value is **5 or more**, *round up* by increasing your desired place value by one and replacing all the digits to the right of it with zeros. (If you are rounding to a place value to the right of the decimal point, you can just drop the remaining digits to the right of it.)

A clever way to recall the rules for rounding is to remember the two phrases "4 or less, let it rest" and "5 or more, let it soar!" These will remind you that if the number to the right of the place value you are rounding to is 0, 1, 2, 3, or 4, then you will keep the desired place value the same, but if that number to the right of the place value you are rounding to is 5, 6, 7, 8, or 9, then you will bump your desired place value digit up by one.

Example

Round 8.963 to the nearest hundredth.

1. Find the hundredths place: 6 is currently in the hundredths place.
2. Investigate the next digit *to the right* of the hundredths place: 3 is in the thousandths place.
3. The digit to the right of the hundredths place (3), is 4 or less, so round down by keeping your last desired digit (6) and drop all the following digits since they are to the right of the decimal: 8.963 rounds to 8.96.

Here's an instance of when you will be rounding to a place that is to the left of the decimal.

Example

Round 342,803 to the nearest thousand.

1. Find the thousands place: 2 is currently in thousands.
2. Investigate the next digit *to the right* of the thousands place: 8 is in the hundreds.
3. The digit to the right of the thousands place (8), is 5 or more, so round up by increasing your last desired digit (2)

by one and replacing all the digits to the right of it with zeros: 342,803 rounds to 343,000.

A special case in rounding up is when your last desired digit is a 9. When that is the case, replace the 9 with a zero and increase the digit to the *left* of the 9 by one. (This is like increasing the 9 to 10.) For example, to round 3,498 to the nearest ten, the 8 in the units place rounds the 9 in the tens place up to 0 and the 4 in the hundreds place up to 5: 3,498 rounded to the nearest 10 is 3,500. (If you think about it, 3,498 is clearly closer to 3,500 than it is to 3,400 or 3,490.)

⚠ CAUTION

Remember, when rounding to the nearest *tenth*, you are looking at the *hundredth* place, and *only* the hundredth place, to see whether it will round up or down. For example, when rounding 0.749 to the nearest tenth, you only look at the 4 in the hundredths place to round it to 0.7, and you ignore the 9 in the thousandths place. Don't make the mistake of looking two digits past your desired rounding place value since this would cause you to make the mistake of rounding 0.749 to 0.75 and then maybe up to 0.8. This is not correct since 0.749 is closer to 0.7 than it is to 0.8.

Practice

You regularly work the closing shift at a card store in the mall. The opening shift employee is not very good at math, so at the end of some days the cash register is a little short on money and other days it has a little more money than it should. When you close the store each night the surplus or deficit of money is recorded. Your boss has asked you to round each day's errors to the nearest dollar, so round each decimal here to the nearest whole number.

_____ **49.** Monday: + $0.26

_____ **50.** Tuesday: – $0.74

_____ **51.** Wednesday: – $9.49

_____ **52.** Thursday: + $3.38

_____ **53.** Friday: – $8.50

_____ **54.** Saturday: + $7.82

_____ **55.** Sunday: – $16.09

Given the following numbers and contexts, name the place value you think makes most sense to round to, and then round each number.

_____ **56.** Julie ran 151.27 miles in May.

_____ **57.** The population of Summit County, Colorado is 27,493.

_____ **58.** A tumor measures 2.37182 microns.

_____ **59.** The starting salary of a post office employee is $51,392.

_____ **60.** Kauai's average rainfall is 68.62 inches.

_____ **61.** The U.S. budget deficit is $128,742,609,517.

Round each decimal to the specified place.

_____ **62.** 8.49; tenths

_____ **63.** 12.875; nearest penny (What place value represents pennies?)

_____ **64.** 87,299,985; hundred thousands

_____ **65.** 549.895; hundreds

_____ **66.** 125.46; nearest whole number

_____ **67.** 86.44; tens

_____ **68.** 9.9999; thousandths

TRY THIS

As you pay for things throughout the day, take a look at the prices. Are they written in dollars and cents? If so, how would you read the numbers aloud using the terms discussed in this lesson? For a bit of a challenge, insert a zero in the tenths column of the number, thereby pushing the two numbers right of the decimal place one place to the right. Now how would you say the amount out loud? Additionally, as you buy items while shopping and as you get your paychecks, notice how you probably automatically round numbers up or down to estimate your expenses or your income.

Answers

Practice Problems

1. One tenth
2. One hundredth
3. One thousandth
4. One ten-thousandth
5. One hundred-thousandth
6. Five and nineteen hundredths
7. One and five hundred twenty-one ten-thousandths
8. Ten and one millionth
9. 0.6 (or .6)
10. 0.06
11. 0.025
12. 0.321
13. 9.006
14. 3.0001
15. 15.216
16. 1.00000101
17. $\frac{1}{10}$
18. $\frac{3}{100}$
19. $\frac{3}{4}$
20. $\frac{99}{100}$
21. $\frac{1}{200}$
22. $\frac{1}{8}$
23. $\frac{23}{500}$
24. $5\frac{1}{25}$
25. $4\frac{3}{20}$
26. $123\frac{9}{20}$
27. $20\frac{1}{200}$
28. $10\frac{2,001}{20,000}$
29. 0.4

30. 0.7
31. $0.1\overline{6}$ *or* $0.16\frac{2}{3}$
32. $0.\overline{714285}$
33. $0.\overline{4}$
34. 4.75
35. $3.\overline{285714}$
36. 16.25
37. $17.8\overline{3}$ or $17.83\frac{1}{3}$
38. $33.\overline{6}$
39. 0.009, 0.05, 0.2
40. 0.396, 0.417, 0.422
41. 0.01, 0.019, 2.009
42. 0.2, 0.28, 0.8, 0.82
43. All have the same value
44. 0.005, 0.05, 0.5, 0.505
45. 1.1, 1.1000001, 1.101, 10.001
46. $0.0\overline{23}$, 0.230, $0.\overline{23}$, 2.3
47. 0.01, 0.011, $\frac{1}{10}$, $\frac{11}{100}$, $0.\overline{1}$
48. 0.05, $\frac{480}{1{,}000}$, 0.49, $\frac{1}{2}$, $\frac{55}{100}$
49. + $0
50. – $1
51. – $9
52. + $3
53. – $9
54. + $8
55. – $16
56. Tens place: 150 miles
57. Thousands place: 27,000 people
58. Hundredths place: 2.37 microns
59. Thousands place: $51,000
60. Tens: 70 inches
61. Billion: $129 billion ($129,000,000,000)
62. 8.49
63. 12.86

64. 87,300,000
65. 500
66. 125
67. 90
68. 1.00

Sample Question 1

1. Write 2 as the whole number: 2
2. Write 7 as the top of the fraction: $2\frac{7}{}$
3. Since there is only one digit to the right of the decimal, it's tenths.
4. Write 10 as the bottom number: $2\frac{7}{10}$

Sample Question 2

1. Set up the division problem: $5\overline{)1}$
2. Add a decimal point and a zero to the divisor (1): $5\overline{)1.0}$
3. Bring the decimal point up into the answer: $5\overline{)1.0}$
4. Divide:

$$\begin{array}{r} 0.2 \\ 5\overline{)1.0} \\ 10 \\ 0 \end{array}$$

Thus, $\frac{1}{5}$ = 0.2, or 2 tenths.

Sample Question 3

1. Since 0.0111 has the greatest number of decimal places (4), tack zeros onto the ends of the other decimals so they all have 4 decimal places:	0.1000, 0.1100, 0.1010, 0.0111
2. Ignore the decimal points and compare the whole numbers:	1,000; 1,100; 1,010; 111
3. The low-to-high sequence of the whole numbers is:	111; 1,000; 1,010; 1,100
Thus, the low-to-high sequence of the original decimals is:	0.0111; 0.1, 0.101, 0.11

CHAPTER

7 Adding and Subtracting Decimals

What would life be without arithmetic, but a scene of horrors?

—SYDNEY SMITH, English writer (1771–1845)

CHAPTER SUMMARY

This second decimal chapter focuses on addition and subtraction of decimals. It concludes by teaching you how to add or subtract decimals and fractions together.

Adding and subtracting decimals comes up all the time, especially when dealing with money. Have you ever been in a store where the cash register has gone out of service and the sales associates must calculate customers' sales tax and correct change on their own? Being comfortable with operations on decimals is important for your performance in the workplace as well as for your own personal use in everyday life. This chapter covers the steps for adding and subtracting decimals and fractions, and also offers some word problems for you to practice your skills in real-world context.

Adding Decimals

Which of the following correctly illustrates how to add 2,030 and 40?

```
2,030    2,030      2,030
+ 40      + 40   + 40
```

You probably recognize that the middle solution is the correct way to add 2,030 and 40, but can you use your new knowledge of decimals and place value to explain *why*? The 4 in 40 is in the tens place, so when adding it to 2,030, it must be lined up with the 3 in the tens place of 2,030.

An easy way to make sure you are adding correctly is to always line your decimal points up directly over each other. If one of your numbers does not have a decimal point, remember to just add it to the end of the number. Here are the rules for adding decimals:

1. Line the numbers up in a column so their decimal points are aligned.
2. Tack zeros onto the ends of shorter decimals to keep the digits lined up evenly.
3. Move the decimal point directly down into the answer area and add as usual.

CAUTION

The number one pitfall in adding and subtracting decimals happens when the numbers are lined up, instead of the **decimals**. Before you start adding or subtracting with decimals, add zeros after the last digit to the right of the decimal point to all the numbers until they each have the same amount of digits to the right of the decimal point. For a whole number, just add a decimal point and then add zeros to the right of it. Example:

Incorrect alignment:	Correct alignment:
4.2	4.200
0.34	0.340
5.871	5.871
+ 18	18.000
	28.411

Example

3.45 + 22.1 + 0.682

1. Line up the numbers so their decimal points are aligned:
   3.45
   22.1
   0.682
2. Tack zeros onto the ends of the shorter decimals to fill in the "holes":
   3.450
   + 22.100
   + 0.682
3. Move the decimal point directly down into the answer area and add:
   + 26.232

To check the *reasonableness* of your work, estimate the sum by using the rounding technique you learned in Lesson 6. Round each number you added to the nearest whole number, and then add the resulting whole numbers. If the sum is close to your answer, your answer is in the ballpark. Otherwise, you may have made a mistake in placing the decimal point or in the adding. Rounding 3.45, 22.1, and 0.682 gives you 3, 22, and 1, respectively. Their sum is 26, which is *reasonably* close to your actual answer of 26.232. Therefore, 26.232 is a *reasonable* answer.

Look at an example that adds decimals and whole numbers together. Remember: A whole number is understood to have a decimal point to its right.

Example

0.6 + 35 + 0.0671 + 4.36

1. Put a decimal point at the right of the whole number (35) and line up the numbers so their decimal points are aligned:

```
  0.6
 35.
  0.0671
  4.36
```

2. Tack zeros onto the ends of the shorter decimals to fill in the "holes":

```
 +0.6000
 35.0000
  0.0671
+ 4.3600
```

3. Move the decimal point directly down into the answer area and add:

```
+40.0271
```

Now you try this sample question. Step-by-step answers to sample questions are at the end of the lesson.

Sample Question 1

A lab in Chicago has e-mailed the following measurements to your lab, recording the growth of a bacteria colony over 4 weeks. Find their sum:

$12 + 0.1 + 0.02 + 0.943$

Practice

The intern at your office added all the following decimals but forgot to add the decimal points. Use estimating to determine where the decimal point should be placed in each of the following sums.

_____ **1.** $3.5 + 3.7 = 72$

_____ **2.** $1.4 + 0.8 = 22$

_____ **3.** $1.79 + 0.21 = 200$

_____ **4.** $4.13 + 2.07 + 5.91 = 1211$

_____ **5.** $4.835 + 1.217 = 6052$

_____ **6.** $9.32 + 4.1 = 1342$

_____ **7.** $7.42 + 125.931 = 133351$

_____ **8.** $6.1 + 0.28 + 4 = 1038$

Add the following decimals.

_____ **9.** 1.789 + 0.219

_____ **10.** 1.48 + 0.9

_____ **11.** 3.59 + 6

_____ **12.** 10.7 + 8.935

_____ **13.** 6.1 + 0.2908 + 4

_____ **14.** 14.004 + 0.9 + 0.21

_____ **15.** 1.03 + 2.5 + 40.016

_____ **16.** 5.2 + 0.7999 + 0.0001

_____ **17.** 0.1 + 0.01 + 0.001 + 1.0001

_____ **18.** 5.1 + 1.05 + 0.510 + 15

Subtracting Decimals

When subtracting decimals, follow the same steps as in adding to ensure that you're subtracting the correct digits and that the decimal point ends up in the right place.

Example
4.8731 – 1.7

1. Line up the numbers so their decimal points are aligned:

    4.8731
    1.7

2. Tack zeros onto the end of the shorter decimal to fill in the "holes": 4.8731
   1.7000
3. Move the decimal point directly down into the answer and subtract: 3.1731

Subtraction is easily checked by adding the number that was subtracted to the difference (the answer). If you get back the other number in the subtraction problem, then your answer is correct. For example, let's check our answer to the previous subtraction problem.

Here's the subtraction:

$$\begin{array}{r} 4.8731 \\ -\underline{1.7000} \\ 3.1731 \end{array}$$

1. Add the number that was subtracted (1.7000) to the difference (3.1731): + 1.7000
2. The subtraction is correct because we got back the first number in the subtraction problem (4.8731): 4.8731

You can check the *reasonableness* of your work by estimating: Round each number to the nearest whole number and subtract. Rounding 4.873 and 1.7 gives 5 and 2, respectively. Since their difference of 3 is close to your actual answer, 3.1731 is *reasonable.*

TIP

Sometimes subtracting mixed decimals will completely eliminate the existing whole numbers and you will just be left with a decimal value less than 1. When this is the case, write a zero in the ones place. Example: 5.67 – 4.9 = **0**.77, and should not be written as .77.

Borrowing

Next, look at a subtraction example that requires borrowing. Notice that borrowing works exactly the same as it does when you're subtracting whole numbers.

Example

2 – 0.456

1. Put a decimal point at the right of the whole number (2) and line up the numbers so their decimal points are aligned:

 2.
 0.456

2. Tack zeros onto the end of the shorter decimal to fill in the "holes":

 2.000
 0.456

3. Move the decimal point directly down into the answer and subtract after borrowing:

 9 9
 1 ~~10~~ ~~10~~ 10
 2.000
 – 0.456
 1.544

4. Check the subtraction by addition:

 1.544
 + 0.456

 Our answer is correct because we got back the first number in the subtraction problem:

 2.000

Sample Question 2

You had $78 in your wallet before you paid $0.78 for a pack of gum. How much money do you now have?

Practice

Subtract the following decimals.

_____ **19.** 6.4 – 1.3

_____ **20.** 1.89 – 0.37

_____ **21.** 12.35 – 8.05

_____ **22.** 2.35 – 0.9

_____ **23.** 5 – 3.81

_____ **24.** 3.2 – 1.23

_____ **25.** 1 – 0.98765

_____ **26.** 2.4 – 2.3999

_____ **27.** 0.02001 – 0.009999

_____ **28.** 1.1111 – 0.88889

Combining Addition and Subtraction

The best way to solve problems that combine addition and subtraction is to "uncombine" them; separate the numbers to be added from the numbers to be subtracted by forming two columns. First create a column where you can find the sum of all numbers that are being added. Then create a column where you can find the sum of all numbers that are being subtracted. Lastly, subtract these two sums to find your final answer.

Example

0.7 + 4.33 – 2.46 + 0.0861 – 1.2

1. Line up the numbers to be *added* so their decimal points are aligned:

$$\begin{array}{r} 0.7 \\ 4.33 \\ 0.0861 \end{array}$$

2. Tack zeros onto the ends of the shorter decimals to fill in the "holes":

$$\begin{array}{r} 0.7000 \\ 4.3300 \\ \underline{+\ 0.0861} \end{array}$$

3. Move the decimal point directly down into the answer and add: 5.1161

4. Line up the numbers to be *subtracted* so their decimal points are aligned:

$$\begin{array}{r} 2.46 \\ 1.2 \end{array}$$

5. Tack zeros onto the end of the shorter decimal to fill in the "holes":

$$\begin{array}{r} 2.46 \\ \underline{+\ 1.20} \end{array}$$

6. Move the decimal point directly down into the answer area and add: 3.66

7. Subtract the step 6 answer from the step 3 answer, lining up the decimal points, filling in the "holes" with zeros, and moving the decimal point directly down into the answer area:

$$\begin{array}{r} 5.1161 \\ \underline{-\ 3.6600} \\ 1.4561 \end{array}$$

Sample Question 3

It turns out the lab in Chicago had some errors in the data they sent to your lab. This is the current data that models the growth and decline of a bacteria colony over several weeks. Add and subtract the following:

$12 + 0.1 - 0.02 + 0.943 - 2.3$

Add and subtract the following decimals.

_____ **29.** $6.4 - 1.3 + 1.2$

_____ **30.** $8.7 - 3.2 + 4$

_____ **31.** $5.48 + 0.448 - 0.24$

_____ **32.** $7 - 0.3 - 3.1 + 3.8$

_____ **33.** $4.7 + 2.41 - 0.8 - 1.77$

_____ **34.** $1 - 0.483 + 3.17$

_____ **35.** $14 - 0.15 + 0.8 - 0.2$

_____ **36.** $22.2 - 3.3 - 4.4 - 5.5$

_____ **37.** $1.111 - 0.8989 + 0.0819 - 0.000009$

_____ **38.** $2.00002 - 0.8881 - 0.99918$

Word Problems

Word problems 39 through 46 involve decimal addition, subtraction, and rounding.

_____ **39.** At the supermarket, Alberto purchased 2.3 pounds of tomatoes, 1.1 pounds of lettuce, a 0.6 pound cucumber, and 4 pounds of carrots. He also decided to buy a 10-pound bag of potatoes. How many pounds of produce did Alberto buy?

_____ **40.** Mia just began a job at a flower shop, where she makes arrangements for large events. During her first week she averaged 8.65 minutes per arrangement and during her second week she got a bit faster and each arrangement took her 7.8 minutes. Express in decimal form how much she improved per arrangement.

_____ **41.** Sidney rang up a customer for a purchase and the total with tax was $70.28. The customer gave Sidney $75. How much change must Sidney give to her customer?

_____ **42.** Chad is in charge of keeping track of and paying the FedEx account at Selmer's Pet Land each month. The balance due at the beginning of September was $278.24. On September 10th he shipped some exotic bird food to a client for $135.30, on the 26th he shipped an aquarium filtration system for $221.28, and on September 30th he made a payment of $450 to the FedEx account. What will his new balance be on October 1st?

_____ **43.** Julie bought three books for $12.95 each, two magazines for $4.09 each, and an annual membership to the VIP Bookhaven Members Club for $122. Estimate her total cost before tax by using rounding.

_____ **44.** Matt bought some used records and CDs that cost $6.99, $12.49, $8.50, and $11.89. The cashier told him the total was $49.87. Was this reasonable? Why or why not?

_____ **45.** Xander walked 1.4 miles to work, a total of 0.7 miles (round-trip) to and from lunch, 0.8 miles each way to UPS to send a proposal to a client, and then 1.4 miles back home. His roommate Lucy ran 4.5 miles at the local track for exercise. Who covered more distance and by how much?

_____ **46.** Brand A's road bike weighs 22.5 pounds, and brand B's bike weighs 22.15 pounds. Which one is lighter? By how much?

Working with Decimals and Fractions Together

When a problem contains both decimals and fractions, you should change the numbers to the same type, either decimals or fractions, depending on which you're more comfortable working with. Consult Lesson 6 if you need to review changing a decimal into a fraction and vice versa.

Example

$\frac{3}{8} + 0.37$

Fraction-to-decimal conversion:

1. Convert $\frac{3}{8}$ to its decimal equivalent:

$$\begin{array}{r} 0.375 \\ 8\overline{)3.000} \\ \underline{24} \\ 60 \\ \underline{56} \\ 40 \\ \underline{40} \\ 0 \end{array}$$

2. Add the decimals after lining up the decimal points and filling the "holes" with zeros:

$$\begin{array}{r} 0.375 \\ \underline{+\,0.370} \\ 0.745 \end{array}$$

Decimal-to-fraction conversion:

1. Convert 0.37 to its fraction equivalent: $\frac{37}{100}$

2. Add the fractions after finding the least common denominator:

$$\begin{array}{r} \frac{37}{100} = \frac{74}{200} \\ +\ \frac{1}{10} = \frac{75}{200} \\ \frac{149}{200} \end{array}$$

Both answers, 0.745 and $\frac{149}{200}$, are correct. You can easily check this by converting the fraction to the decimal or the decimal to the fraction.

Practice

Add and subtract these decimals and fractions.

_____ **47.** $\frac{1}{2} + 0.5$

_____ **48.** $\frac{1}{4} + 0.25$

_____ **49.** $\frac{5}{8} + 0.5$

_____ **50.** $4.9 + \frac{3}{10}$

_____ **51.** $\frac{3}{20} + 2.6$

_____ **52.** $3.15 + 2\frac{3}{4}$

_____ **53.** $2.75 + \frac{5}{12}$

_____ **54.** $1.11 - 1\frac{1}{10} - 0.01$

_____ **55.** $\frac{11}{5} - 2.5 + \frac{3}{10}$

_____ **56.** $\frac{1}{2} + 0.2 + 2.01 - \frac{1}{5}$

TRY THIS

Now that you know how to add and subtract decimals, when you go shopping, see whether you are able to determine how much change you will get from the cashier when you pay for a single item with a $5, $10, or $20 bill. See whether you can correctly calculate your change *before* the cashier rings you up. Once you are able to calculate your change correctly, see whether you can keep a running estimate of how much a multi-item purchase will cost. How close can you get to the total cost of an entire basket of groceries? Use rounding and see how good you can be at keeping a running total of your purchases.

Answers

Practice Problems

1. 7.2
2. 2.2
3. 2.00 or 2
4. 12.11
5. 6.052
6. 13.42
7. 133.351
8. 10.38
9. 2.008
10. 2.38
11. 9.59
12. 19.635
13. 10.3908
14. 15.114
15. 43.546
16. 6
17. 1.1111
18. 21.66
19. 5.1
20. 1.52
21. 4.3
22. 1.45
23. 1.19
24. 1.97
25. 0.01235
26. 0.0001
27. 0.010011
28. 0.22221
29. 6.3
30. 9.5
31. 5.688

32. 7.4
33. 4.54
34. 3.687
35. 14.45
36. 9 or 9.0
37. 0.293991
38. 0.11274
39. 18 pounds
40. 0.85 minutes
41. $4.72
42. $184.82
43. $164
44. No, this is not reasonable because when you round all the decimals to whole numbers the sum is $40 and since the cashier's total is almost $50, there must be an error.
45. Xander covered 0.6 miles more than Lucy.
46. Brand B, by 0.35 pounds
47. 1
48. 0.5 or $\frac{1}{2}$
49. 1.125 or $1\frac{1}{8}$
50. 5.2 or $5\frac{1}{5}$
51. 2.75 or $2\frac{3}{4}$
52. 5.9 or $5\frac{9}{10}$
53. $3.1\overline{6}$ or $3\frac{1}{6}$
54. 1
55. 0
56. 2.51 or $2\frac{51}{100}$

Sample Question 1

1. Line up the numbers and fill the "holes" with zeros, like this:

$$\begin{array}{r} 12.000 \\ 0.100 \\ 00.020 \\ +\ 0.943 \\ \hline \end{array}$$

2. Move the decimal point down into the answer and add: 13.063

Sample Question 2

1. Line up the numbers and fill the "holes" with zeros, like this:

$$\begin{array}{r} 78.00 \\ -\ 0.78 \\ \hline \end{array}$$

2. Move the decimal point down into the answer and subtract: 77.22
3. Check the subtraction by adddition: It's correct: You got back the first number in the problem.

$$\begin{array}{r} 77.22 \\ +\ 0.78 \\ \hline 78.00 \end{array}$$

Sample Question 3

1. Line up the numbers and fill the "holes" with zeros, like this:

$$\begin{array}{r} 12.000 \\ 0.100 \\ +\ 0.943 \\ \hline \end{array}$$

2. Move the decimal point down into the answer and add: 13.043
3. Line up the numbers and fill the "holes" with zeroes:
4. Move the decimal point down into the answer and add:

$$\begin{array}{r} 0.02 \\ +\ 2.30 \\ \hline 2.32 \end{array}$$

5. Subtract the sum of step 4 from the sum of step 2, after lining up the decimal points amd filling the "holes" with zeroes:

$$\begin{array}{r} 13.043 \\ -\ 2.320 \\ \hline 10.723 \end{array}$$

6. Check the subtraction by adddition: It's correct: You got back the first number in the problem.

$$\begin{array}{r} 10.723 \\ +\ 2.320 \\ \hline 13.043 \end{array}$$

CHAPTER

8 Multiplying and Dividing Decimals

Can you do division? Divide a loaf by a knife—what's the answer to that?

—From *Through the Looking Glass*, by Lewis Carroll, English author and mathematician (1832–1898)

CHAPTER SUMMARY

This final decimal chapter focuses on multiplication and division.

If you need to order 8.5 pounds of chicken for a catered event and the chicken costs $4.89 per pound, you are going to need to perform decimal multiplication to get the total cost. When you are calculating sales tax, mortgage interest rates, or discounts offered by stores, decimal multiplication and division is required. Some of the most useful and important skills to master in this unit are the shortcuts applied to multiplying and dividing by multiples of 10.

Multiplying Decimals

To multiply decimals:

1. Ignore the decimal points and multiply as you would whole numbers.
2. Count the number of decimal digits (the digits to the *right* of the decimal point) in both of the numbers you multiplied. Do NOT count zeros tacked onto the end as decimal digits.
3. Beginning at the right side of the product (the answer), count left that number of digits, and put the decimal point to the left of the last digit you counted.

Example

1.57 3 2.4

1. Multiply 157 times 24:

$$\begin{array}{r} 157 \\ \times\ 24 \\ \hline 628 \\ 314 \\ \hline 3768 \end{array}$$

2. Because there are a total of three decimal digits in 1.57 and 2.4, count off 3 places from the right in 3768 and place the decimal point to the *left* of the third digit you counted (7): 3.768

When multiplying with decimals it is easy to make a careless error with the final placement of your decimal point. Therefore, it is very important to always use the rounding technique presented in Lesson 6 to estimate your answer so that you can compare it to your final answer to see how reasonable your solution is. Round each number you multiplied to the nearest whole number, and then multiply the results. If the product is close to your answer, your answer is in the ballpark. Otherwise, you may have made a mistake in placing the decimal point or in multiplying. Rounding 1.57 and 2.4 to the nearest whole numbers gives you 2 and 2. Their product is 4, which is close to your answer. Thus, your actual answer of 3.768 seems reasonable.

CAUTION

The number of decimal digits in 3.768000 is three, NOT six.

Now you try. Remember, step-by-step answers to sample questions are at the end of the lesson.

Sample Question 1

You are purchasing 2.7 feet of chain that costs $3.26 per foot. First use rounding to estimate what the total cost will be, and then use multiplication to find the exact total cost.

In multiplying decimals, you may get a product that doesn't have enough digits for you to put in a decimal point. In that case, tack zeros onto the left of the product to give your answer enough digits; then add the decimal point.

Example

0.03×0.006

1. Multiply 3 times 6: $3 \times 6 = 18$
2. The answer requires 5 decimal digits because there are a total of

five decimal digits in 0.03 and 0.006. Because there are only 2 digits in the answer (18), tack three zeros onto the left: 00018

3. Put the decimal point at the front of the number (which is 5 digits in from the right): .00018

Sample Question 2

Onions are on sale for $0.40/pound and you want to buy one onion that weighs 0.2 of a pound. Use multiplication to find the cost of this onion and then check your answer by rounding.

Practice

Multiply these decimals.

_____ **1.** 0.01×0.6

_____ **2.** 3.1×4

_____ **3.** 0.1×0.20

_____ **4.** 15×0.210

_____ **5.** 0.875×8

_____ **6.** 78.2×0.0412

Multiplication Shortcut

Although it is important to know how to perform decimal multiplication by hand, it is likely that on the job, you will often use a calculator to perform the types of multiplication problems you just practiced. The most useful and important multiplication skill is to be able to perform the mental math of quickly multiplying numbers by 10 and by multiples of 10 (like 20, 80, and 1,400). Look at these tricks and see whether you can recognize the pattern:

- To multiply a number by 10, move the decimal point **one digit to the right.**
- To multiply a number by 100, move the decimal point **two digits to the right.**
- To multiply a number by 1,000, move the decimal point **three digits to the right.**

Are you seeing the pattern? In general, just count the number of zeros the multiple of ten has, and move the decimal point that same number of spaces to the **right.** If you don't have enough digits, you will need to add zeros onto the right side of your answer.

Example

1,000 × 3.82

1. Since there are three zeros in 1,000, move the decimal point in 3.82 three digits to the right.
2. Since 3.82 has only two decimal digits to the right of the decimal point, add one zero on the right before moving the decimal point: 3.820

Thus, 1,000 × 3.82 = 3,820

As mentioned earlier, this multiplication shortcut can be used with all multiples of 10. When multiplying by any number that ends with a zero (or zeros), you can ignore the zeros at first, perform the multiplication, and add the zeros onto your answer at the end. How does this apply to $220 \times 3{,}000$? Ignore the zeros at first and think of this problem as $22 \times 3 = 66$. Since you ignored four zeros, you can conclude that $220 \times 3{,}000 = 660{,}000$.

Example

2,500 employees of Tina's Tinkerland will get a $30 Coffe Haus gift card as a holiday gift. Perform 2,500 × $30 quickly for your boss to see what this will cost.

1. Think of 2,500 × $30 as 25 × 3, which equals 75.
2. You ignored three zeros, so adding them back on you get, 2,500 × $30 = $75,000.

Practice

Use the shortcuts for multiplying by multiples of 10 to perform the following decimal multiplication:

_____ **7.** 300×700

_____ **8.** 10×3.6400

_____ **9.** 100×0.01765

_____ **10.** $1{,}000 \times 38.71$

_____ **11.** $450 \times 2{,}000$

_____ **12.** $9{,}000 \times 4{,}000$

Dividing Decimals

Dividing Decimals by Whole Numbers

To divide a decimal by a whole number, bring the decimal point straight up into the answer (the *quotient*), and then divide as you would normally divide whole numbers.

Example

$4\overline{)0.512}$

1. Move the decimal point straight up into the *quotient* area: $4\overline{)0\dot{\uparrow}512}$

2. Divide:

$$\begin{array}{r} 0.128 \\ 4\overline{)0.512} \\ \underline{4} \\ 11 \\ \underline{8} \\ 32 \\ \underline{32} \\ 0 \end{array}$$

3. To check your division, multiply the quotient (0.128) by the *divisor* (4).

$$\begin{array}{r} 0.128 \\ \underline{\times \quad 4} \\ 0.512 \end{array}$$

If you get back the *dividend* (0.512), you know you divided correctly.

Sample Question 3

A patient must receive 0.125 milligrams of medicine, evenly divided over 5 days. Perform the division 0.125 ÷ 5 to see how much he should receive in each daily dosage.

Dividing by Decimals

To divide any number by a decimal, first change the problem into one in which you're dividing by a whole number.

1. Move the decimal point to the right of the number by which you're dividing (the *divisor*).
2. Move the decimal point the same number of places to the right in the number you're dividing into (the *dividend*).
3. Bring the decimal point straight up into the answer (the *quotient*) and divide.

Remember that dividing by a decimal that has a value less than 1 will give you a bigger number!

Example

$0.03\overline{)1.215}$

1. Because there are two decimal digits in 0.03, move the decimal point two places to the right in both numbers: $0.03.\overline{)1.21.5}$
2. Move the decimal point straight up into the quotient: $3.\overline{)121.5}$
3. Divide using the new numbers:

$$\begin{array}{r} 40.5 \\ 3\overline{)121.5} \\ \underline{12} \\ 01 \\ \underline{00} \\ 15 \\ \underline{15} \\ 0 \end{array}$$

Under the following conditions, you'll have to tack zeros onto the right of the last decimal digit in the dividend, the number into which you're dividing:

Case 1. There aren't enough digits to move the decimal point to the right.

Case 2. The answer doesn't come out evenly when you divide.

Case 3. You're dividing a whole number by a decimal. In this case, you'll have to tack on the decimal point as well as some zeros.

Case 1

There aren't enough digits to move the decimal point to the right.

Example

$0.5\overline{)1.2}$

1. Because there are two decimal digits in 0.03, the decimal point must be moved two places to the right in both numbers. Since there aren't enough decimal digits in 1.2, tack a zero onto the end of 1.2 before moving the decimal point: $0.\underset{\smile\smile}{03}.\overline{)1.\underset{\smile\smile}{20}.}$
2. Move the decimal point straight up into the quotient: $3.\overline{)120\overset{\cdot}{\uparrow}}$
3. Divide using the new numbers:

$$
\begin{array}{r}
40. \\
3\overline{)120.} \\
\underline{12} \\
00 \\
\underline{00} \\
0
\end{array}
$$

Case 2

The answer doesn't come out evenly when you divide.

Example

0.5)1.2

1. Because there is one decimal digit in 0.5, the decimal point must be moved one place to the right in both numbers — 0.5.)1.2.
2. Move the decimal point straight up into the quotient: 5.)12.
3. Divide, but notice that the division doesn't come out evenly:

```
  2.
5)12.
  10
   2
```

4. Add a zero to the end of the dividend (12.) and continue dividing:

```
   2.4
5)12.0
  10
   20
   20
    0
```

Example

0.3).10

1. Because there is one decimal digit in 0.3, the decimal point must be moved one place to the right in both numbers: 0.3.).1.0
2. Move the decimal point straight up into the quotient: 3.)1.0

3. Divide, but notice that the division doesn't come out evenly:

$$\begin{array}{r} 0.3 \\ 3\overline{)12.0} \\ \underline{9} \\ 1 \end{array}$$

4. Add a zero to the end of the dividend (1.0) and continue dividing:

$$\begin{array}{r} 0.33 \\ 3\overline{)1.00} \\ \underline{9} \\ 10 \\ \underline{9} \\ 1 \end{array}$$

5. Since the division still did not come out evenly, add another zero to the end of the dividend (1.00) and continue dividing:

$$\begin{array}{r} 0.333 \\ 3\overline{)1.000} \\ \underline{9} \\ 10 \\ \underline{9} \\ 10 \\ \underline{9} \\ 1 \end{array}$$

6. By this point, you have probably noticed that the quotient is a repeating decimal. Thus, you can stop dividing and write the quotient like this: $0.\overline{3}$

Case 3

When you're dividing a whole number by a decimal, you have to tack on the decimal point as well as some zeros.

Example

$0.02\overline{)19}$

1. There are two decimals in 0.02, so we have to move the decimal point to the right two places in both numbers. Because 19 is a whole number, put its decimal point at the end (19.), add two zeros to the end (19.00), and then move the decimal point to the right twice (1900.): $0.02.\overline{)19.00.}$
2. Move the decimal point straight up into the quotient: $2.\overline{)1900.}$
3. Divide using the new numbers:

$$
\begin{array}{r}
950 \\
2\overline{)1900} \\
\underline{18} \\
10 \\
\underline{10} \\
00 \\
\underline{00} \\
0
\end{array}
$$

Sample Question 4

Kern County uses benzalkonium chloride to treat the water in their farm-raised fish bed. They have 3 gallons left and must put 0.06 gallons of benzalkonium chloride in the fish bed every day to keep it algae free. See how many days they can treat their fish bed with their remaining 3 gallons of benzalkonium chloride by performing the division 3 ÷ 0.06.

Practice

Divide.

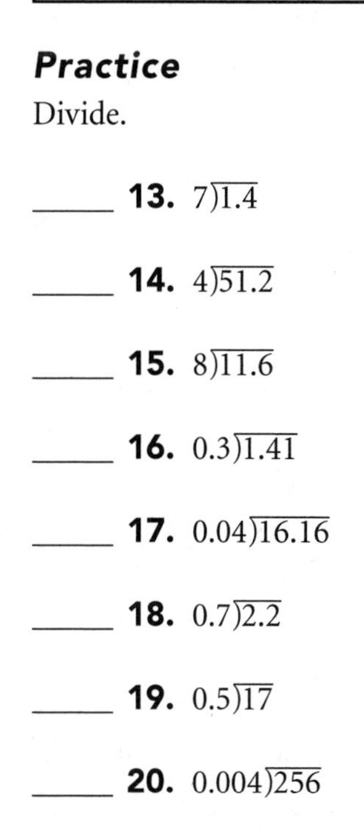

_____ **13.** $7\overline{)1.4}$

_____ **14.** $4\overline{)51.2}$

_____ **15.** $8\overline{)11.6}$

_____ **16.** $0.3\overline{)1.41}$

_____ **17.** $0.04\overline{)16.16}$

_____ **18.** $0.7\overline{)2.2}$

_____ **19.** $0.5\overline{)17}$

_____ **20.** $0.004\overline{)256}$

Division Shortcut

The multiplication shortcut given for multiples of 10 can be reversed for division with multiples of 10. Look at these tricks and see whether you can recognize the pattern:

- To divide a number by 10, move the decimal point **one digit to the left**.

- To divide a number by 100, move the decimal point **two digits to the left.**
- To divide a number by 1,000, move the decimal point **three digits to the left.**

Hopefully now that you've already learned the trick for multiplication, the pattern for division is easy to see. Count the number of zeros the multiple of ten has, and move the decimal point that same number of spaces to the **left**. If you don't have enough digits, you will need to add zeros onto the left side of your answer before you add the decimal point.

Example

Divide 12.345 by 1,000.

1. Since there are three zeros in 1,000, move the decimal point in 12.345 three digits to the left.
2. Since 12.345 only has two digits to the left of its decimal point, add one zero at the left, and then move the decimal point: 0.**0**12.345

Thus, 12.345 ÷ 1,000 = 0.012345

Practice

Perform the following division problems using the division shortcut.

_____ **21.** 450,000 ÷ 1,000

_____ **22.** 2,348 ÷ 100

_____ **23.** 0.07 ÷ 10

_____ **24.** 3.14 acres of land divided among 100 farmers

_____ **25.** $4,000,000 of international aid money divided among 100,000 refugees

_____ **26.** 700 pounds of grain divided among 1,000 chickens

TIP

When solving problems in the workplace and in your everyday life, sometimes the addition of a decimal place can make it confusing to choose correctly between multiplication and division. For example, if I tell you that 5 people are going to share 20 pieces of candy, it is probably easy for you to see that each person will get 4 pieces of candy, since 20 ÷ 5 = 4. However, if you need to create 5.5 servings out of 20.83 pounds of potatoes, it is a little harder to see that the division problem of 20.83 ÷ 5.5 will determine how big each serving will be. In order to simplify decimal word problems, *remove* the decimals and think about how you would solve the problem with whole numbers, and then use that operation with the original decimals.

Decimal Word Problems

The following are word problems involving decimal multiplication and division.

_____ **27.** Luis earns $7.25 per hour. Last week, he worked 37.5 hours. How much money did he earn that week, rounded to the nearest cent?

_____ **28.** At $6.50 per pound, how much do 2.75 pounds of cookies cost, rounded to the nearest cent?

_____ **29.** Anne drove a truckload of merchandise to the mall, averaging 40.2 miles per hour for 1.6 hours. How many miles did she drive rounded to the nearest tenth of a mile?

_____ **30.** Jordan built a total of 12.4 bookshelves in 4 days. On average, how many bookshelves did he build each day?

_____ **31.** One almond contains 0.07 milligrams of iron. Nurse Fontanesi wants her patient to get 14 milligrams of iron per day. If her patient got her iron only through almonds how many would she need to eat each day to get 14 milligrams?

_____ **32.** If Chedder cheese costs $4.00 a pound, how many pounds can you get for $2.50?

_____ **33.** Suzy Sattler makes, sells, and ships Valentine's Day chocolates to her customers all around the country. If each individual box of candy is 2.5 inches tall, how many high can she stack in a shipping package that is 35 inches tall?

_____ **34.** Mark has 27 stacks of dimes, each of which consists of 7 dimes. How much money does he have?

TRY THIS

If you make an hourly wage:

Write down how much money you earn per hour, in dollars and cents. (If you don't have a job yet, make up an hourly wage for yourself and make it generous!) First multiply your wage by 40 hours to see how much money you earn per week. Now multiply that answer by 50 to see how much you could earn in a year if you took 2 weeks of vacation. Next, divide your hourly wage by 60 to see how much you earn per minute. Which answers make sense to round to the nearest penny and which make sense to round to the nearest dollar?

If you make a weekly or monthly salary:

Divide your salary by the number of hours you work per week or per month to see how much you make per hour. Now divide your hourly wage by 60 to see how much you earn per minute. Now give yourself a raise by adding $0.50 to your hourly salary and see how that would increase your weekly or monthly salary. Does it make more sense for you to round this final answer to the nearest penny or whole dollar?

Answers

Practice Problems

1. 0.006
2. 12.4
3. 0.02
4. 3.15
5. 7 or 7.000
6. 3.22184
7. 210,000
8. 36.4
9. 1.765
10. 38,710
11. 90,000
12. 36,000,000
13. 0.2
14. 12.8
15. 1.45
16. 4.7
17. 404
18. $3.\overline{142857}$
19. 34
20. 64,000
21. 450
22. 23.48
23. 0.007
24. 0.0314 acres per farmer
25. $40 of international aid money per refugee
26. 0.7 pounds of grain per chicken
27. $271.88
28. $17.88
29. 64.3 miles
30. 3.1 bookshelves

31. 200 almonds
32. 0.625 pounds
33. 14 boxes
34. $18.90

Sample Question 1

1. First estimate the product by rounding 3.26 down to 3 and 2.7 up to 3:

$3 \times 3 = 9$, so your answer should be close to $9.

2. Multiply 326 times 27:

$$\begin{array}{r} 326 \\ \times\, 27 \\ \hline 2282 \\ +652 \\ \hline 8802 \end{array}$$

3. Because there are a total of three decimal digits in 3.26 and 2.7, count off three places from the right in 8802 and place the decimal point to the left of the third digit you counted (8): 8.802

4. Since 8.802 rounds to the nearest penny as 8.80, the total cost will be $8.80. $8.80 is a reasonable answer since your original estimate with rounding was $9.

Sample Question 2

1. Multiply 4 times 2:

$$\begin{array}{r} 4 \\ \times 2 \\ \hline 8 \end{array}$$

2. The answer requires two decimal digits. Because there is only one digit in the answer (8), tack one zero onto the left: 08

3. Put the decimal point at the front of the number (which is two digits in from the right): .08

4. Reasonableness check: Round both numbers to the nearest whole number and multiply: $0 \times 0 = 0$, which is reasonably close to your answer of 0.08.

Sample Question 3

1. Move the decimal point straight up into the quotient: $5\overline{)0.125}$

2. Divide:

$$\begin{array}{r} 0.025 \\ 5\overline{)0.125} \\ \underline{0} \\ 12 \\ \underline{10} \\ 25 \\ \underline{25} \\ 0 \end{array}$$

3. Check: Multiply the quotient (0.025) by the divisor (5):
Since you got back the dividend (0.125), the division is correct:

$$\begin{array}{r} 0.025 \\ \times \quad 5 \\ \hline 0.125 \end{array}$$

Sample Question 4

1. Because there are two decimal digits in 0.06, the decimal point must be moved two places to the right in both numbers. Since there aren't enough decimal digits in 3, tack a decimal point and two zeros onto the end of 3 before moving the decimal point: $.06.\overline{)3.00.}$

2. Move the decimal point straight up into the quotient: $6.\overline{)3.00.}$

3. Divide using the new numbers:

$$\begin{array}{r} 50. \\ 6.\overline{)300.} \\ \underline{30} \\ 00 \\ \underline{00} \\ 0 \end{array}$$

4. Check: Multiply the quotient (50) by the original divisor (0.06):

$$\begin{array}{r} 50 \\ \underline{\times\ 0.60} \\ 3.00 \end{array}$$

Since you got back the original dividend (3), the division is correct:

CHAPTER

9 Working with Percents

Mathematics is a language.

—Josiah Willard Gibbs, theoretical physicist (1839–1903)

CHAPTER SUMMARY

This first percent chapter is an introduction to the concept of percents. It explains the relationships between percents, decimals, and fractions.

Percents are all around us in the workplace, in the news, and in our personal lives. Maybe you need to enlarge a logo by 200% for an advertisement you're creating, or maybe you can get a 10% discount for purchasing office supplies in bulk, or maybe you are comparing mortgages that offer either a 5% fixed interest rate or a 4% variable interest rate. Like decimals and fractions, you need to understand and feel comfortable working with percents on the job if you want to be a valuable employee. In this lesson you'll be learning what percentages are and how to change percentages into equivalent forms. In the next lesson, you'll learn why percentages are so useful and how to apply them at work and at home.

What, exactly, is a percent? Let's break down the word to see. The prefix, "per," means "for every," as in, "Melissa makes $12 per hour." The root, "cent," means "100," as can be recognized in the words *century*, *cent*, and *centipede*. Put these two parts together and the word *percent* means "for every 100." Therefore, it should not comes as a shock for you to hear that a *percent* is a special kind of fraction that is *out of 100*. Knowing this, we can conclude that 5% means "5 for every 100" or "5 out of 100." This is $\frac{5}{100}$, which is the way to represent 5% as a fraction. Using what we know about fractions and decimals, we know that $\frac{5}{100}$ is five hundredths, which is 0.05. Therefore, 0.05 is the way to represent 5% as a decimal.

In summary, percents can be expressed in two different ways:

1. As a fraction (by putting the number over 100): $5\% = \frac{5}{100}$
2. As a decimal (by moving the decimal point two places to the left): $5\% = 0.05$

The three different ways to write the same number, as a percent, fraction, or decimal are all illustrated in the triangle, and this type of equivalence triangle can be made for *all* percents.

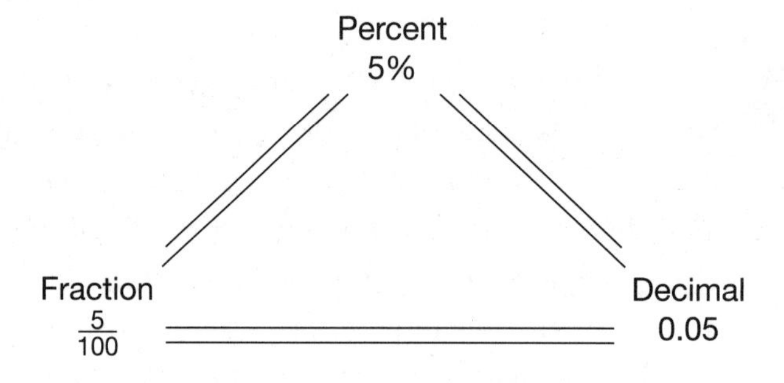

As you'll learn in the following box, it is necessary to convert percents into either their equivalent fraction or decimal forms *before* using them in calculations. If you are shaky on fractions or decimals you will want to review the previous lessons before reading further.

⚠ CAUTION

Percentages are only used in writing and are never used in calculations. In order to perform algebraic calculations with percentages, they must first be converted to decimals or fractions. This is because 5% is really a shortcut to saying "5 out of 100" or writing $\frac{5}{100}$. Notice that the percent symbol (%) looks like a jumbled up "100"; let that remind you to change a percentage into a fraction (by putting it over 100), or into its decimal equivalent (by moving the decimal 2 spaces to the left), before doing any calculations. For example, when needing to take 40% of a number, you will not use "40" in your calculations, but you will instead use $\frac{40}{100}$ or 0.40.

Changing Percents to Decimals

Hopefully after reading the preceding box you understand why it is crucial for you to be able to change percentages into decimal or fraction format.

To change a percent to a decimal, drop the percent sign and move the decimal point two digits to the **left**. Remember: If a number doesn't have a decimal point, it's assumed to be at the right. If there aren't enough digits to move the decimal point, add zeros on the **left** before moving the decimal point.

Example

Change 20% to a decimal.

1. Drop the percent sign:	20
2. There's no decimal point, so put it at the right:	20.
3. Move the decimal point two digits to the left:	0.20.

Thus, 20% is equivalent to 0.20, which is the same as 0.2. (Remember: Zeros at the right of a decimal don't change its value.)

Now you try these sample questions. The step-by-step solutions are at the end of this lesson.

Sample Questions 1 and 2

Change 75% to a decimal.

Change 142% to a decimal.

Practice

Change these percents to decimals.

_____ **1.** 1%

_____ **2.** 19%

_____ **3.** 0.001%

_____ **4.** 4.25%

_____ **5.** 0.04%

_____ **6.** $1\frac{1}{4}\%$

_____ **7.** $87\frac{1}{2}\%$

_____ **8.** 150%

_____ **9.** 9.999%

_____ **10.** 10,000%

Changing Percents to Fractions

Rather than changing a percentage into a decimal in order to perform algebraic operations, it is sometimes easier to do a particular calculation with a fraction. Therefore, being able to write percentages as fractions is another useful skill.

To change a percent to a fraction, remove the percent sign and write the number over 100; then reduce if possible.

Example

Change 20% to a fraction.

1. Remove the % and write the fraction 20 over 100: $\frac{20}{100}$
2. Reduce: $\frac{20 \div 20}{100 \div 20} = \frac{1}{5}$

Example

Change $16\frac{2}{3}\%$ to a fraction.

1. Remove the % and write the fraction $16\frac{2}{3}$ over 100: $\frac{16\frac{2}{3}}{100}$

2. Since a fraction means "top number divided by bottom number," rewrite the fraction as a division problem: $16\frac{2}{3} \div 100$
3. Change the mixed number ($16\frac{2}{3}$) to an improper fraction ($\frac{50}{3}$): $\frac{50}{3} \div \frac{100}{1}$
4. Flip the second fraction ($\frac{100}{1}$) and multiply: $\frac{\overset{1}{\cancel{50}}}{3} \times \frac{1}{\underset{2}{\cancel{100}}} = \frac{1}{6}$

Sample Question 3

Change $33\frac{1}{3}$% to a fraction.

Practice

Change these percents to fractions.

_____ **11.** 3%

_____ **12.** 25%

_____ **13.** 0.03%

_____ **14.** 60%

_____ **15.** 3.75%

_____ **16.** 37.5%

_____ **17.** $87\frac{1}{2}$%

_____ **18.** 110%

_____ **19.** $250\frac{1}{4}\%$

_____ **20.** 1,000%

Changing Decimals to Percents

Our focus in this chapter is to become proficient at conversions between percentages, fractions, and decimals and in the next chapter we'll learn why percentages are so useful. You are halfway there! Next let's look at decimal to percent conversions.

To change a decimal to a percent, move the decimal point two digits to the **right**. If there aren't enough digits to move the decimal point, add zeros on the right before moving the decimal point. If the decimal point moves to the very right of the number, don't write the decimal point. Finally, tack on a percent sign (%) at the end.

Example

Change 0.3 to a percent.

1. Move the decimal point two digits to the right after adding one zero on the right so there are enough decimal digits: 0.30.
2. The decimal point moved to the very right, so remove it: 30
3. Tack on a percent sign: 30%

Thus, 0.3 is equivalent to 30%.

CAUTION

$0.3 \neq 3\%$. Remember that $0.3 = 0.30$, which is equivalent to $\frac{30}{100}$, and this is why $0.3 = 30\%$ and not 3%. Don't forget to move the decimal point *two* digits to the right.

TIP

When changing decimals to percentages, remember that a mixed decimal is always going to be more than 100%. Example: 1 = 100% and 2.75 = 275%

Sample Questions 4 and 5

Change 0.875 to a percent.

Change 0.7 to a percent.

Change these decimals to percents.

_____ **21.** 0.85

_____ **22.** 0.9

_____ **23.** 0.02

_____ **24.** 0.008

_____ **25.** 0.031

_____ **26.** 0.667

_____ **27.** 2.5

_____ **28.** 1.25

_____ **29.** 100

_____ **30.** 5,000.1

Changing Fractions to Percents

Changing fractions to percents is one of the most important conversions with percents because it allows us to easily compare and represent data that is presented as ratios, such as "3 out of 5 employees don't use all their vacation days every year." You will learn all about this in the next lesson, but for now let's look at how this converting from fractions to percents is done.

To change a fraction to a percent, there are two techniques. Each is illustrated by changing the fraction $\frac{1}{5}$ to a percent.

Technique 1

1. Multiply the fraction by 100.

2. Multiply $\frac{1}{5}$ by 100: $\frac{1}{\cancel{5}_1} \times \frac{\cancel{100}^{20}}{1} = 20\%$

Note: Change 100 to its fractional equivalent, $\frac{100}{1}$, before multiplying.

Do you recall that decimals are turned into percentages by moving the decimal place two places to the right and adding a percent symbol? (0.35 becomes 35%.) Well, moving the decimal place twice is the same thing as multiplying it by 100. This is the method you just learned in Technique 1 for converting a fraction into a percent. Remember that whether you are changing a decimal *or* a fraction into a percent, you will be multiplying by 100.

Technique 2

1. Divide the fraction's bottom number into the top number; then move the decimal point two digits to the **right** and tack on a percent sign (%).
2. Divide 5 into 1, move the decimal point 2 digits to the right, and tack on a percent sign: $5\overline{)1.00}$ = 0.20 → 0.20. → 20%

Note: You can get rid of the decimal point because it's at the extreme right of 20.

Sample Question 6

Change $\frac{1}{9}$ to a percent.

Changing Mixed Numbers to Percents

To change a mixed number to a percent, first change it to an improper fraction and then, apply either Technique 1 or Technique 2.

Example

Change $2\frac{3}{8}$ to a percent.

1. Change $2\frac{3}{8}$ to an improper fraction: $\frac{19}{8}$
2. Applying Technique 2, divide the bottom number into the top number; then, move the decimal point two digits to the right and tack on a percent sign: $8\overline{)19.000}$ = 2.375 $\rightarrow$ 237.5%

So, $2\frac{3}{8} = 237.5\%$.

Change these fractions to percents.

_____ **31.** $\frac{1}{2}$

_____ **32.** $\frac{1}{6}$

_____ **33.** $\frac{19}{25}$

_____ **34.** $\frac{7}{4}$

_____ **35.** $\frac{18}{5}$

_____ **36.** $\frac{5}{8}$

_____ **37.** $1\frac{1}{25}$

_____ **38.** $3\frac{3}{4}$

_____ **39.** $10\frac{1}{10}$

_____ **40.** $9\frac{2}{3}$

Common Equivalences of Percents, Fractions, and Decimals

There are certain percentages that we see in the workplace, in stores, and in real-world settings more than others. Since you need to change those percentages into either decimals or fractions before using them in calculations, it's a good idea to memorize the fraction-decimal-percentage equivalences in the following table. If your boss wants you to sell 25% of your inventory by the end of the week, it's helpful to quickly know that he wants $\frac{1}{4}$ of your inventory out the door. Pay special attention to the fraction column and notice how the fractions have all been reduced.

CONVERTING DECIMALS, PERCENTS, AND FRACTIONS

DECIMAL	PERCENT	FRACTION
0.25	25%	$\frac{1}{4}$
0.5	50%	$\frac{1}{2}$
0.75	75%	$\frac{3}{4}$
0.1	10%	$\frac{1}{10}$
0.2	20%	$\frac{1}{5}$

0.4	40%	$\frac{2}{5}$
0.6	60%	$\frac{3}{5}$
0.8	80%	$\frac{4}{5}$
$0.\overline{3}$	$33\frac{1}{3}\%$	$\frac{1}{3}$
$0.\overline{6}$	$66\frac{2}{3}\%$	$\frac{2}{3}$

Practice

After memorizing the table, cover up any two columns with a piece of paper and write the equivalences. Check your work to see how many numbers you remembered correctly. Do this exercise several times, with sufficient time between to truly test your memory.

TRY THIS

Find out what your local sales tax is. (Some places have a sales tax of 3% or 6.5%, for example.) Convert that percentage into a fraction and reduce it to its lowest terms. Then, go back to the original sales tax percentage and convert it into a decimal. Now you'll be able to recognize your sales tax no matter what form it's written in. Think about the actual meaning of the sales tax you pay. If your sales tax is 8%, what does that mean you are being charged for every 100 pennies ($1), that an item costs? Using this information, can you estimate how much the tax will be on a $1 pen, a $2 card, and a $5 pair of socks?

Answers

Practice Problems

1. 0.01
2. 0.19
3. 0.00001
4. 0.0425
5. 0.0004
6. 0.0125
7. 0.875
8. 1.50
9. 0.09999
10. 100
11. $\frac{3}{100}$
12. $\frac{1}{4}$
13. $\frac{3}{10{,}000}$
14. $\frac{3}{5}$
15. $\frac{3}{80}$
16. $\frac{3}{8}$
17. $\frac{7}{8}$
18. $1\frac{1}{10}$
19. $2\frac{5{,}025}{10{,}000}$ or $2\frac{201}{400}$
20. 10
21. 85%
22. 90%
23. 2%
24. 0.8%
25. 3.1%
26. 66.7%
27. 250%
28. 125%
29. 10,000%

30. 500,010%
31. 50%
32. $16.\overline{6}\%$ or $16\frac{2}{3}\%$
33. 76%
34. 175%
35. 360%
36. 62.5% or $62\frac{1}{2}\%$
37. 104%
38. 375%
39. 1,010%
40. $966\frac{2}{3}\%$ or $966.\overline{6}\%$

Sample Question 1

1. Drop the percent sign:	75	142
2. There's no decimal point, so put one at the right:	75.	142.
3. Move the decimal point two digits to the left:	0.75.	1.42

Thus, 75% is equivalent to 0.75 and 142% is equivalent to 1.42.

Sample Question 2

1. Move the decimal point two digits to the right:	0.87.5
2. Tack on a percent sign:	87.5%

Thus, 0.875 is equivalent to 87.5%.

Sample Question 3

Technique 1:

1. Multiply $\frac{1}{9}$ by 100%: $\frac{1}{9} \times \frac{100}{1}\% \; 5 \; \frac{100}{9}\%$

2. Convert the improper fraction ($\frac{100}{9}$) to a decimal: $\frac{100}{9} = 11.\overline{1}\%$

Or, change it to a mixed number: $\frac{100}{9}\% = 11\frac{1}{9}\%$

Thus, $\frac{1}{9}$ is equivalent to both $11.\overline{1}\%$ and $11\frac{1}{9}\%$.

Technique 2:

1. Divide the fraction's bottom number (9) into the top number (1):

```
  0.111 etc.
9)1.000 etc.
  0
  10
   9
   10
    9
    10
```

2. Move the decimal point in the quotient two digits to the **right** and tack on a percent sign (%): $11.\overline{1}\%$

Note: $11.\overline{1}\%$ is equivalent to $11\frac{1}{9}\%$.

Sample Question 4

1. Move the decimal point two digits to the right after tacking on a zero: 0.70.
2. Remove the decimal point because it's at the extreme right: 70
3. Tack on a percent sign: 70%

Thus, 0.7 is equivalent to 70%.

Sample Question 5

1. Remove the % and write the fraction $33\frac{1}{3}$ over 100: $\frac{33\frac{1}{3}}{100}$
2. Since the fraction means "top number divided by bottom number," rewrite the fraction as a division problem: $33\frac{1}{3} \div 100$
3. Change the mixed number $33\frac{1}{3}$ to an improper fraction ($\frac{100}{3}$): $\frac{100}{3} \div \frac{100}{1}$
4. Flip the second fraction $\frac{100}{1}$ and multiply: $\frac{\cancel{100}^{1}}{3} \times \frac{1}{\cancel{100}_{1}} = \frac{1}{3}$

Thus, $33\frac{1}{3}$ is equivalent to the fraction $\frac{1}{3}$.

CHAPTER

10 Introduction to Percent Word Problems

There is no branch of mathematics, however abstract, which may not someday be applied to the phenomena of the real world.

—Nicolai Lobachevsky, Russian mathematician (1792–1856)

CHAPTER SUMMARY

This second percent chapter focuses on the three main varieties of percent word problems and some real-life applications.

Knowing how to convert a given percent into a decimal or fraction demonstrates a basic level of understanding of percents. The ultimate goal is to be able to use percentages to represent and understand real-word data in the workplace in order to help analyze information. Percentages can be helpful indicators of patterns and trends, which are important factors in making many business decisions. There are a very wide variety of applications of how percents are used to interpret data, but let's start with the following example, which illustrates how percentages can be more useful than raw data.

The table compares trials of how successfully Drug A and Drug B brought headache relief in 15 minutes or less. Can you say with certainty which drug had better results?

HEADACHE RELIEF IN 15 MINUTES OR LESS	NUMBER OF PEOPLE REPORTING SUCCESS	TOTAL NUMBER OF PEOPLE IN TRIAL
Drug A	470	1,250
Drug B	448	1,120

Was Drug A better since it helped 22 more people than Drug B? This would not be an accurate way of reading the results since Drug A had more people in its trial—therefore, it isn't correct to merely compare the number of successes. It is possible to compare the drugs by looking at the fractions $\frac{470}{1,250}$ and $\frac{448}{1,120}$, but these are very ugly fractions.

Now compare the same information represented in the next table using percentages.

HEADACHE RELIEF IN 15 MINUTES OR LESS	PERCENT OF PEOPLE REPORTING SUCCESS
Drug A	37.6%
Drug B	40.0%

This table makes it easy to quickly see that Drug B had better success than Drug A. This is one example of how percentages are key in comparing data sets of different sizes. In this lesson you are going to learn about the three basic types of applied percentage problems. In the next lesson you'll learn about percentage increase and decrease as well as some other percentage techniques.

The percent skills you will first master apply to these three different types of percent applications:

1. **Finding a percent of a whole number.**
   - Your boss asked you to mark down a rack of merchandise by 20%, so you need to find 20% of all the different prices in order to put the items on sale.
2. **Finding what percent a number is of another number.**
   - Your manager asked you to poll customers in the supermarket to see how many preferred nonfat milk to low-fat milk. You found that 14 out of the 58 people you surveyed preferred nonfat milk. What percent is 14 out of 58?
3. **Finding the whole when the percent and part are given.**
   - If 15% of a shipment of eggs arrived broken, and 30 eggs arrived broken, how many eggs were in the total shipment?

The One-Size-Fits-All Approach to Percent Problems

While each variety has its own approach, there is a single shortcut formula you can use to solve each of these.

Here, this shortcut is written in two different ways—you can remember the one that makes the most sense to you. (Some people prefer using "*part* to *whole*" and others find it easier to remember the phrase "*is* over *of*.")

$$\frac{\text{part}}{\text{whole}} = \frac{\%}{100} \text{ OR } \frac{\text{is}}{\text{of}} = \frac{\%}{100}$$

part The number that usually follows (but can precede) the word ***is*** in the question.

whole The number that usually follows the word ***of*** in the question.

% The number in front of the % or word ***percent*** in the question.

To solve each of the three main varieties of percent questions, use the fact that the cross products are equal. The **cross products** are the products of the quantities diagonally across from each other. Remembering that *product* means *multiply*, here's how to create the cross products for the percent shortcut:

$$\frac{part}{whole} \times \frac{\%}{100}$$

$$part \times 100 = whole \times \%$$

When working with equations like the previous one, where one quantity is equal to another quantity, it is important to remember these two tips for solving for the missing variable:

1. Use opposite operations to get the missing variable alone. (Addition and subtraction are opposites, and multiplication and division are opposites.) Therefore, if the unknown quantity in an equation is being *multiplied* by a number, *divide* by that number in order to get the unknown by itself.
2. When working with equations, whatever you do to one side you *must* also do to the other side to get the correct answer. So if you divide the left side by 5, you must also divide the right side by 5.

Finding a Percent of a Whole

You are finding the percent of a whole number in each of these scenarios: when you calculate the 20%-off sale prices of inventory in your store, when you determine how much money your 3% pay raise will be next year, and when you figure out how much the 9% tax will be for a customer's purchase. Finding the percent of a whole number is one of the most useful skills to have with percentages. To find the percent of a whole number, you are looking for the part, so plug the other given information into the percent shortcut and solve for the part.

Example

Your food bill is $40 and you want to calculate a tip of 15%. What is 15% of 40?

1. 15 is the **%**, and 40 is the ***whole***: $\frac{\textbf{\textit{part}}}{40} = \frac{15}{100}$
2. Cross multiply and solve for ***part***: $\textbf{\textit{part}} \times 100 = 40 \times 15$

$\textbf{\textit{part}} \times 100 = 600$

Thus, **6** ***is*** 15% of 40. $\mathbf{6} \times 100 = 600$

Note: If the answer didn't leap out at you when you saw the equation, you could have divided both sides by 100:

$$\text{part} \times 100 = 600$$

$$\div 100 \quad \div 100$$

$$\text{part} = \frac{600}{100}, \text{ so } \underline{\underline{\text{part} = 6}}$$

TIP

When solving real-life problems that involve percentages, boil the content down to a single, fundamental question. In the previous example question, after reading, "Your food bill is $40 and you want to calculate a tip of 15%," you should say to yourself, "*What is 15% of 40?*" Ignore all the descriptive words and distractions and identify exactly what it is you need to find.

Example

20% of the 25 students in Mr. Mann's class failed the test. How many students failed the test? (Hint: The fundamental question here is "What is 20% of 25?")

1. The ***percent*** is 20 and the ***whole*** is 25, since it follows the word ***of*** in the problem.

$$\frac{\textbf{\textit{part}}}{25} = \frac{20}{100}$$

2. Cross multiply and solve for ***part***:

$$\textbf{\textit{part}} \times 100 = 25 \times 20$$
$$\textbf{\textit{part}} \times 100 = 500$$
$$\mathbf{5} \times 100 = 500$$

Thus, 5 students failed the test. Again, if the answer doesn't leap out at you, divide both sides of the equation by 100, leaving ***part*** = 5.

Now you try finding the percent of a whole with the following sample question. The step-by-step solution is at the end of this lesson.

Sample Question 1

90% of the 300 dentists surveyed recommended sugarless gum for their patients who chew gum. How many dentists did NOT recommend sugarless gum?

Practice

Find the percent of the whole number in each given situation:

_____ **1.** 60% of 500 people surveyed are unhappy with their Internet service provider.

_____ **2.** 20% off a $320 air compressor

_____ **3.** 9.5% sales tax on a $15,000 piece of equipment

_____ **4.** 0.2% interest earned on a bank account with a $6,300 balance

_____ **5.** 25% collections fee for a $220 bill that was never paid

_____ **6.** A 12-inch sketch needs to be replicated to 240% of its original size.

_____ **7.** A tumor that was 0.6 microns is now 150% of its original size.

_____ **8.** 110% of an investment that was originally $70,000

Finding What Percent a Number Is of Another Number

Finding what percent one number is of another number is another skill that is widely applicable to real-world math in the workplace and at home. When you survey 58 customers and only 14 of them prefer nonfat milk, you can express that finding as a percent. Or when you tell your boss that you have exceeded the average sales in your office for 16 out of the past 18 weeks, it is a more convincing figure when expressed as a percentage. To find what percent one number is of another number, you are comparing the *part* to the *whole* and you can put these numbers into the percent shortcut and then solve for the percent.

Example

Phil's Flooring Service bid on 40 different projects for floor repair work in September, but only 10 of those estimates resulted in signed contracts. What percentage of bids resulted in contracts? (Hint: The fundamental question here is "10 is what percent of 40?")

1. 10 is the ***part*** and 40 is the ***whole***:

$$\frac{10}{40} = \frac{\%}{100}$$

2. Cross multiply and solve for %:

$$10 \times 100 = 40 \times \%$$
$$1{,}000 = 40 \times \%$$
$$1{,}000 = 40 \times \mathbf{25}$$

Thus, 10 is **25%** of 40.

If you weren't sure where to go from $1{,}000 = 40 \times \%$, recall that in order to get the % by itself, you can use opposite operations and divide both sides by 40:

$$1{,}000 = 40 \times \%$$
$$\div 40 \quad \div 40$$
$$\frac{480}{1{,}000} = \%, \text{ so } \underline{\underline{\% = 25}}$$

TIP

When solving for percent using the percent shortcut, the percent symbol is over 100, so the percent is being written as a percent and not as a decimal. This means that you do not move your decimal point after you have solved for percent with this shortcut, since it is already being represented as "out of 100." Simply add the percent symbol to your final answer.

Example

35 members of the 105-member marching band are girls. What percent of the marching band is girls?

1. The ***whole*** is 105 because it follows the word ***of*** in the problem: Therefore, 35 is the ***part*** because it is the other number in the problem, and we know it's not the percent because that's what we must find:

$$\frac{35}{105} = \frac{\%}{100}$$

$$35 \times 100 = 105 \times \%$$

$$3{,}500 = 105 \times \%$$

2. Divide both sides of the equation by 105 to find out what % is equal to:

$$3{,}500 = 105 \times \%$$
$$\underline{\div 105 \quad \div 105}$$

$$\frac{3{,}500}{105} = \%, \text{ so } \% = 33\frac{1}{3}$$

Thus, $33\frac{1}{3}\%$ of the marching band is girls.

Sample Question 2

The quality control step at the Light Bright Company has found that 2 out of every 1,000 light bulbs tested are defective. Assuming that this batch is indicative of all the light bulbs they manufacture, what percent of the manufactured light bulbs is defective?

Practice

Find the percentage represented by each situation.

_____ **9.** 16 out of 50 people in Yardly bike to work.

_____ **10.** 24 out of 300 people refused to participate in the survey.

_____ **11.** 4 out of 800 calculators had a defect.

_____ **12.** A batter hit 3 out of the 12 pitches he swung for in the last game.

_____ **13.** A pianist has 12 out of the 18 pages of music memorized.

_____ **14.** Haley spent $22.82 of her $32.60 tip money on gas.

_____ **15.** A professional athlete weighs 165 lbs. and only 3.3 lbs. of that is fat.

_____ **16.** Kaoru's bank account with a $5,600 balance earned just $8.40 last year in interest.

Finding the Whole When the Percent and Part Are Given

Although this particular type of problem is the least common of the three presented in this lesson, at times it is necessary to be able to work backwards by using the percent and the part to figure out what the whole is. The good news is, this type of problem is no more difficult than the others and can still be solved with the percent shortcut.

Example

As part of a fundraiser, Gina's boss offered to give 40% of the pizza place's lunch sales to a local dog shelter. With the first order of the afternoon, Gina tells him they've already raised $20 for the dog shelter. How much was the entire order for? (Hint: 20 is 40% of what number?)

1. 20 is the ***part*** and 40 is the %: $\frac{20}{\textit{\textbf{whole}}} = \frac{40}{100}$
2. Cross multiply and solve for the ***whole***:

$$20 \times 100 = \textit{\textbf{whole}} \times 40$$
$$2{,}000 = \textit{\textbf{whole}} \times 40$$

Thus, 20 is 40% ***of* 50.** $\quad 2{,}000 = 50 \times 40$

Note: To solve the last step, you could divide both sides by 40 to get **whole** alone:

$$2{,}000 = \text{whole} \times 40$$
$$\div 40 \qquad \div 40$$
$$\frac{2{,}000}{40} = \text{whole, so } \underline{\underline{\text{whole} = 50}}$$

Example

John left a $3 tip, which was 15% of his bill. How much was his bill?

In this problem, $3 is the ***part***, even though there's no ***is*** in the actual question. You know this for two reasons: (1) It's the ***part***

John left for his server, and (2) the word ***of*** appears later in the problem: *of the bill*, meaning that the amount ***of*** the bill is the ***whole***. And, obviously, 15 is the **%** since the problem states ***15%***.

So, here's the setup and solution:

$$\frac{3}{whole} = \frac{15}{100}$$
$$3 \times 100 = \boldsymbol{whole} \times 15$$
$$300 = \boldsymbol{whole} \times 15$$
$$300 = 20 \times 15$$

Thus, John's bill was $20.

Note: Some problems may ask you a different question. For instance, what was the total amount that John's lunch cost? In that case, the answer is the amount of the bill **plus** the amount of the tip, or $23 ($20 + $3).

Sample Question 3

The combined city and state sales tax in Bay City is $8\frac{1}{2}$%. The Bay City Boutique collected $600 in sales tax for sales on May 1. What was the total sales figure for that day, excluding sales tax?

Practice

Find the whole that is represented by each situation.

_____ **17.** The 12% hotel tax was $78. How much did the reservation cost before tax?

_____ **18.** Siena left a 20% tip of $36. How much did the meal cost without the tip?

_____ **19.** A toddler's 45% weight increase over the past year was 14 pounds. What was her original weight?

_____ **20.** A CEO invests $12,480 on furniture, which is 32% of the office renovation budget. How much is the entire office renovation budget?

_____ **21.** A 15% increase it productivity resulted in 300 extra toys being produced per week. How many toys were previously being produced per week?

Which Is Bigger, the *Part* or the *Whole*?

In most percent word problems, the ***part*** is smaller than the ***whole***, as you would probably expect. But don't let the size of the numbers fool you: The ***part*** can, in fact, be larger than the ***whole***. In these cases, the percent will be greater than 100%. In fact, you *want* this to be the case when you make an investment. If you invest $500 you hope that in the future the value of your investment will not fall to 80%, but will instead rise to be 120% or even 200% or its original value.

Example

Caleb is pricing a new shipment of socks so that they can be put out on the floor for sale. A package of socks that had a wholesale cost to the store of $5 will get a retail price of $10 for its customers. What percent markup does Caleb's store use for the socks? (Hint: $10 is what percent of $5?)

1. The ***part*** is 10, and the ***whole*** is 5. $\frac{10}{5} = \frac{\%}{100}$
2. Cross multiply and solve for %: $10 \times 100 = 5 \times \%$

$$1{,}000 = 5 \times \%$$

Thus, 10 is **200%** of 5, which is the same as saying that 10 is **twice as big** as 5: $1{,}000 = 5 \times \mathbf{200}$

Example

Jocelyn is an artist who sells her paintings in local coffee shops around Portland. She sells a painting to Killingsworth Koffee for $400 and one of their customers falls in love with it. If Killingsworth Koffee sells this painting to their customer for $900, what percent was this selling price of their original $400 purchase price?

1. This question is asking, "$900 is what percent of $400?" The coffee shop sold it for more than they paid for it, so their markup was above 100%, which means that $400 is the ***whole*** and the $900 is the ***part***: $\frac{900}{400} = \frac{\%}{100}$
2. Cross multiply: $90{,}000 = 400 \times \%$
3. Divide both sides by 400 to solve for %:

$$90{,}000 = 400 \times \%$$
$$\underline{\div 400 \quad \div 400}$$
$$\frac{90{,}000}{400} = \%, \text{ so } \underline{\underline{\% = 225\%}}$$

Therefore, Killingworth Koffee's selling price of $900 was 225% of the $400 they paid Jocelyn for the painting.

Example

Larry gave his taxi driver \$9.20, which included a 15% tip. How much did the taxi ride cost, excluding the tip?

1. The \$9.20 that Larry gave his driver included the 15% tip plus the cost of the taxi ride itself, which translates to: \$9.20 = the cost of the ride + 15% of the cost of the ride

   Mathematically, the cost of the ride is the same as 100% of the cost of the ride, because 100% of any number (like 3.58295) is that number (3.58295). Thus: \$9.20 = 100% of the cost of the ride + 15% of the cost of the ride, or \$9.20 = 115% of the cost of the ride (by addition)

2. \$9.20 ***is*** 115% ***of*** the cost of the ride: $\frac{9.20}{\textbf{\textit{whole}}} = \frac{115}{100}$

   Cross multiply and solve for the ***whole***:

   $$9.20 \times 100 = 115 \times \textbf{\textit{whole}}$$
   $$920 = 115 \times \textbf{\textit{whole}}$$
   $$920 = 115 \times \mathbf{8}$$

   You probably needed to divide both sides by 115 to solve this one. That leaves you with 8 = ***whole***.

   Thus, \$9.20 is 115% of **\$8**, which is the amount of the taxi ride itself.

Sometimes you will know the percentage of a particular outcome, but you need to solve for a quantity regarding the opposite outcome. In cases like these, remember that all the different percentages of a single situation must add up to 100% to represent the entire body of data. Therefore, you can subtract the known percentage from 100% to find the remaining percentage. Consider the following: *70% of people living in Rudabarker, ID prefer dogs over cats as pets. If Rudabarker has a population of 2,800, how many people prefer cats?* Since 70% of the people prefer dogs, that means that 30% prefer cats, since 100% – 70% = 30%. In this case, it is quicker to find 30% of 2,800 instead of finding 70% of 2,800 and then subtracting the number of dog lovers from the entire population.

Percent Word Problems

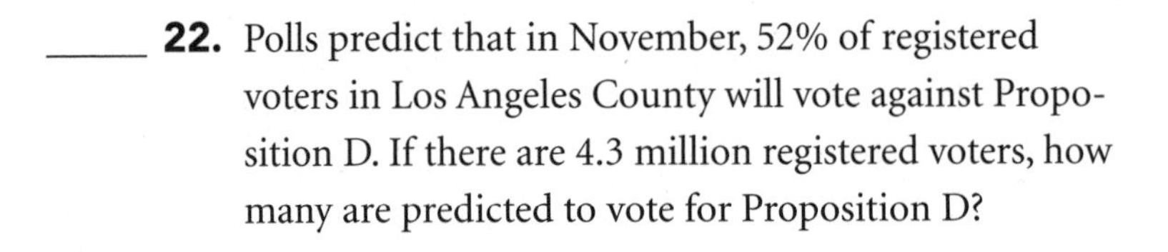

_____ **22.** Polls predict that in November, 52% of registered voters in Los Angeles County will vote against Proposition D. If there are 4.3 million registered voters, how many are predicted to vote for Proposition D?

_____ **23.** A store's New Year's sale is offering 30% off all shoes and bags. If Jo-Anne buys a bag for $80 and a pair of shoes for $120, what will her discount be on her purchase?

_____ **24.** An MP3 player costs $220, and sales tax is $7\frac{3}{4}$%. What will the tax be?

_____ **25.** 134 employees came to an optional employee enrichment seminar in June and 78 of them thought it was a successful event. 180 employees came to the next employee enrichment seminar in August and 95 employees felt it was successful. What percentage of employees felt each of the seminars was successful? Was June or August better received?

_____ **26.** Six and one-half feet of thread are used in one pair of jeans. If 1 foot 11 inches of that thread is used in decorative stitching on the pockets, approximately *what percentage* of thread is used for decorating the pockets?

_____ **27.** An e-mail blast went out to 580 Growing Minds clients with a sales promotion. 104 of these clients acted on the promotion and made an online purchase. In the past, Growing Minds has had an average success rate of 20% purchase activity with these types of online promotions. What was the success rate of this particular promotion, and did it exceed or fall short of previous promotions?

_____ **28.** The company 21st Century Musika earned profit on 5% of the 280 million CDs sold this year. How many CDs does this account for?

_____ **29.** Handle with Care, a nonprofit organization that specializes in educating young parents, spends 7% of its contributions per year on marketing. If it spent $27,120.00 on marketing last year, how much did it earn in contributions? Round to the nearest cent.

_____ **30.** Terry tells the office manager not to book sales appointments during the weekend after Thanksgiving, since people are often out shopping and miss their scheduled appointments. The manager ignores Terry's suggestion and books appointments anyway during the holiday weekend. If 80% of Terry's appointments are not home when he shows up, and only 4 appointments are successfully held that weekend, how many appointments in total did Terry's manager book that weekend?

_____ **31.** Jenn writes for *Rolling Stone*, covering only jazz and country musicians. In her career, 40% of her stories have been about jazz musicians. If she has written 72 stories about country musicians, how many stories has she written about jazz musicians?

_____ **32.** A certain car sells for $20,000, if it is paid for in full (the cash price). However, the car can be financed with a 10% down payment and monthly payments of $1,000 for 24 months. How much more money is paid for the privilege of financing, excluding tax? What percent is this of the car's cash price?

TRY THIS

Whenever you're in a library, on a bus, in a large work area, or any place where there are more than five people gathered together, count the total number of people and write down that number. Then count how many men there are and figure out what percentage of the group is male and what percentage is female. Think of other ways of dividing the group: What percentage is wearing blue jeans? What percentage has black or dark brown hair? What percentage is reading?

Answers

Practice Problems

1. 300 unhappy people
2. $64 off
3. $1,425 in sales tax
4. $12.60 in interest earned
5. $55 collections fee
6. 28.8 inches
7. 0.9 microns
8. $77,000
9. 32% bike to work
10. 8% refused
11. 0.5% defect rate
12. 25% batting success (this is called "250" in baseball circles, which means 0.250 = 25%)
13. $66.\overline{6}$ or $66\frac{2}{3}$ memorized
14. 70% spent on gas
15. 2% body fat
16. 0.15% interest
17. $650 reservation
18. $180 meal
19. 40 lbs.
20. $39,000 budget
21. 2,000 toys
22. 2,236,000 people
23. $60 discount
24. Tax = $17.05
25. June = 58.2%, August = 52.8%, so June was better received
26. 30%
27. 17.9% or 18%, so previous promotions have been more successful

28. 14 million
29. $387,428.57
30. 20 meetings
31. 48
32. $6,000 and 30%

Sample Question 1

There are two ways to solve this problem.

Method 1: Calculate the number of dentists who recommended sugarless gum using the $\frac{part}{whole}$ technique, and then subtract that number from the total number of dentists surveyed to get the number of dentists who did NOT recommend sugarless gum.

1. The ***whole*** is 300, and the **%** is 90: $\frac{part}{300} = \frac{90}{100}$
2. Cross multiply and solve for ***part***: ***part*** × 100 = 300 × 90
   part × 100 = 27,000
   Thus, 270 dentists recommended sugarless gum. **270** × 100 = 27,000
3. Subtract the number of dentists who recommended sugarless gum from the number of dentists surveyed to get the number of dentists who did NOT recommend sugarless gum: 300 – 270 = 30

Method 2: Subtract the percent of dentists who recommended sugarless gum from 100% (reflecting the percent of dentists surveyed) to get the percent of dentists who did NOT recommend sugarless gum. Then, use the $\frac{\textit{part}}{\textit{whole}}$ technique to calculate the number of dentists who did NOT recommend sugarless gum.

1. Calculate the % of dentists who did NOT recommend sugarless gum: $100\% - 90\% = 10\%$
2. The ***whole*** is 300, and the **%** is 10: $\frac{\textit{part}}{300} = \frac{10}{100}$
3. Cross multiply and solve for ***part***: $\textit{part} \times 100 = 300 \times 10$

   $\textit{part} \times 100 = 3{,}000$

   Thus, **30** dentists did NOT recommend sugarless gum. $\mathbf{30} \times 100 = 3{,}000$

Sample Question 2

1. 2 is the ***part*** and 1,000 is the ***whole***: $\frac{2}{1{,}000} = \frac{\%}{100}$
2. Cross multiply and solve for **%**: $2 \times 100 = 1{,}000 \times \%$

   $200 = 1{,}000 \times \%$

   Thus, **0.2%** of the light bulbs are assumed to be defective. $200 = 1{,}000 \times \mathbf{0.2}$

Sample Question 3

1. Since this question includes neither the word ***is*** nor ***of***, it takes a little more effort to determine whether 600 is the ***part*** or the ***whole***! Since $600 is equivalent to $8\frac{1}{2}$% tax, we can conclude that it is the ***part***. The question is asking this: "$600 tax ***is*** $8\frac{1}{2}$% ***of*** what dollar amount of sales?"

   Thus, 600 is the ***part***, and $8\frac{1}{2}$ is the **%**: $\frac{600}{whole} = \frac{8\frac{1}{2}}{100}$

2. Cross multiply and solve for the ***whole***: $600 \times 100 = whole \times 8\frac{1}{2}$

   $60{,}000 = whole \times 8\frac{1}{2}$

   You must divide both sides of the equation by $8\frac{1}{2}$ to get the answer: $60{,}000 \div 8\frac{1}{2} = whole \times 8\frac{1}{2} \div 8\frac{1}{2}$

   $\mathbf{7{,}058.82} = whole$

   Thus, $600 is $8\frac{1}{2}$% of approximately **$7,058.82** (rounded to the nearest cent), the total sales on May 1, excluding sales tax.

Sample Question 2

CHAPTER 11

Another Approach to Percents

The hardest thing in the world to understand is the income tax.

—Albert Einstein, theoretical physicist (1879–1955)

CHAPTER SUMMARY

This final percent chapter introduces another approach to solving percent problems, which some people may find easier than the method covered in the previous chapter. It also gives shortcuts for quickly calculating percentages ending with 5 and 0. Lastly, this lesson will teach you how to calculate percent increase and percent decrease.

Those who prefer a more visual way of solving problems may favor the percent shortcut taught in the past lesson. There is another way of solving percent questions, which is a more algebraic method that others will prefer. This new technique is based on translating a word problem practically *word for word* from

English statements into mathematical symbols and statements. The most important translation rules you'll need to know are:

- ***of*** means multiply (×)
- ***is*** means equals (=)
- ***what*** means a variable, such as w or x

You can apply this direct approach to each of the three main varieties of percent problems.

Finding a Percent of a Whole

Example
What is 15% of 50? (50 is the **whole**.)

Translation

- The word ***What*** is the unknown quantity; use the variable w to stand for it.
- The word ***is*** means *equals* (=).
- Mathematically, ***15%*** is equivalent to both 0.15 and $\frac{15}{100}$ (your choice, depending on whether you prefer to work in decimals or fractions).
- ***of 50*** means *multiply by 50* (× 50).

Using all the information listed here, we translate the problem word by word.

What is 15% of 50?

$$w = \frac{15}{100} \times 50$$

or

$$w = 0.15 \times 50$$

Now solve for w:

$w = 7.5$

Thus, 7.5 is 15% of 50.

The sample questions in this chapter are the same as those in Chapter 10. Solve them again, this time using the direct approach. Step-by-step solutions are at the end of the lesson.

Sample Question 1

Ninety percent of the 300 dentists surveyed recommended sugarless gum for their patients who chew gum. How many dentists did NOT recommend sugarless gum?

Finding What Percent a Number Is of Another Number

Example

10 is what percent of 40?

Translation

- ***10 is*** means *10 is equal to* (10 =).
- ***what percent*** is the unknown quantity, so let's use $\frac{w}{100}$ to stand for it. (The variable w is written as a fraction over 100 because the word *percent* means *per 100*, or *over 100.*)
- ***of 40*** means multiply by 40 (× 40).

Using all the information listed here, we translate the problem word by word.

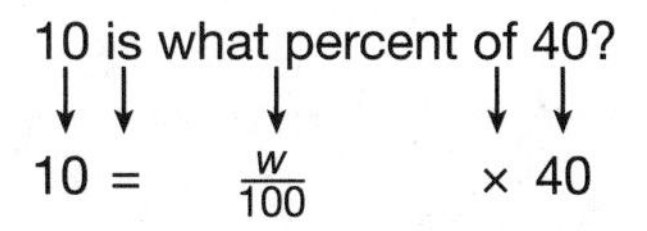

Write 10 and 40 as fractions: $\frac{10}{1} = \frac{w}{100} \times \frac{40}{1}$

Multiply fractions: $\frac{10}{1} = \frac{w \times 40}{100 \times 1}$

Reduce: $\frac{10}{1} = \frac{w \times 2}{5}$

Cross multiply: $10 \times 5 = w \times 2$

Solve by dividing both sides by 2: $25 = w$

Thus, 10 is **25%** of 40.

CAUTION

Since the variable *w* is being written above a 100 denominator, it is being written as a percent and not as a decimal. Therefore, do not move the decimal of your answer—just add the % symbol.

Sample Question 2

The quality-control step at the Light Bright Company has found that 2 out of every 1,000 light bulbs tested are defective. Assuming that this batch is indicative of all the light bulbs they manufacture, what percent of the manufactured light bulbs is defective?

Finding the Whole When a Percent Is Given

Example

20 is 40% of what number?

Translation

- ***20 is*** means *20 is equal to* (20 =).
- Mathematically, ***40%*** is equivalent to both 0.40 (which is the same as 0.4) and $\frac{40}{100}$ (which reduces to $\frac{2}{5}$). Again, it's your choice, depending on which form you prefer.
- ***of what number*** means *multiply by the unknown quantity*; let's use w for it ($\times w$).

Using all the information listed here, we translate the problem word by word.

$$20 = 0.4 \times w$$
$$\underline{\div 0.4 \quad \div 0.4}$$
$$50 = w$$

or

$$20 = \frac{2}{5} \times w$$
$$\frac{20}{1} = \frac{2}{5} \times \frac{w}{1}$$
$$\frac{20}{1} = \frac{2 \times w}{5}$$
$$20 \times 5 = 2 \times w$$
$$100 = 2 \times w$$
$$100 = 2 \times \mathbf{50}$$

Thus, 20 is 40% of **50**.

When using the translation method to solve word problems it is necessary to boil down all the given information into a single essential question before translating, as we began to do in the previous lesson. For example, what would the single essential question be for the following situation: *If Samantha is buying a roll of $80 fabric that is 30% off, what will the discounted, pre-tax, price be?* Using the reasoning presented in an earlier TIP box, you should see that Samantha will be paying 70% of the original price. Therefore the essential question you will use for your translation is "*What is 70% of $80?*"

Sample Question 3

The combined city and state sales tax in Bay City is $8\frac{1}{2}$%. The Bay City Boutique collected $600 in sales tax on May 1. What was the total sale figure for that day, excluding sales tax?

Practice

Use the translating approach to solve the following questions and decide whether you prefer the translating approach or the percent shortcut approach covered in Lesson 10.

_____ **1.** What is 25% of 600 students?

_____ **2.** 14 sick customers is what percent of 80 customers?

_____ **3.** 15 broken pens is 12% of how many pens?

_____ **4.** What is 8.5% tax on a $90 paper order?

_____ **5.** 14 critical injuries is what percent of the 280 passengers in a train accident?

_____ **6.** Ms. Lee's bank account gets 0.2% interest each month. If she gets $0.80 in interest in February, what was her beginning February balance?

The 10%, 15%, & 20% Tipping Shortcut

Have you ever received a bill in a restaurant where you needed to quickly calculate the appropriate tip? In Europe, servers earn a higher hourly wage than in the United States, so it's acceptable to tip 5% there, and a 10% tip is considered *generous*. But here in the United States, it is common to tip 15% to 20% depending on the quality of service you receive. There are tricks to easily calculate all percentages that are multiples of 10% and 5%, which you can use for much more than just figuring out tips. Read on!

1. First, realize that finding 10% of any number is as easy as moving the decimal point one digit to the left:
   Examples:
   - 10% of $35.00 is $3.50.
   - 10% of $82.50 is $8.25.
   - 10% of $59.23 is $5.923, which rounds to $5.92.

   Once you have 10% of a number, you can easily find other percentages that are multiples of 10% such as 20% or 30% or 80%, by multiplying the 10% by the appropriate number:
   - 10% of $40.00 is $4.00
   - 20% = 10% × 2, so 20% of $40.00 is $4.00 × 2 = $8.00
   - 30% = 10% × 3, so 30% of $40.00 is $4.00 × 3 = $12.00
   - 80% = 10% × 8, so 80% of $40.00 is $4.00 × 8 = $32.00

2. Next, calculate 5% of the given number by taking half of the 10% you calculated in Step 1:

   Examples:

   - 5% of $35.00 is half of $3.50, which is $1.75.
   - 5% of $82.50 is half of $8.25, which is $4.125, which rounds to $4.13.
   - 5% of $59.23 is approximately half of $5.92, which is $2.96. (We said approximately because we rounded $5.923 down to $5.92. We're going to be off by a fraction of a cent, but that really isn't significant.)

3. Finally, you can calculate *any* multiple of 5% by correctly manipulating your answers in Step 1 and Step 2. Let's start by finding 15% of our numbers by adding the 10% and 5%:

   Examples:

   - 15% of $35.00 = $3.50 + $1.75 = $5.25
   - 15% of $82.50 = $8.25 + $4.13 = $12.38
   - 15% of $59.23 = $5.92 + $2.96 = $8.88

In an actual situation, you might want to round each calculation **up** to a more convenient amount of money to leave, such as $5.50, $12.50, and $9 if your server was good; or round **down** if your service wasn't terrific.

Example

Find 35% of $120 by using the shortcuts just illustrated:

First calculate 10% and 5% of the $120:	10% of $120 is $12, so 5% of $120 is $6
Next determine how to use 10% and 5% to combine to 35%:	35% = (10% × 3) + (5%)
Lastly, use the answers we just found to calculate 35% of $60:	($12 × 3) + $6 = $42

Sample Question 4

65% of Hamilton County's registered voters showed up to vote on a measure to help increase funding for the school district. The county has 12,013 registered voters. Use rounding and the shortcut just covered to quickly estimate the approximate number of voters who turned up to cast their votes.

Practice

Use the shortcut for multiples of 5% and 10% to find each of the following percentages to the nearest nickel.

_____ **7.** 15% of $25

_____ **8.** 40% of $800

_____ **9.** 20% of $86.40

_____ **10.** 75% of $4,000

_____ **11.** 5% of $18.42

Sale Price Shortcut

There is a convenient shortcut to finding the sale price of items that have a percentage discount. Let's consider a situation where Sam's $50 printer ink is 20% off. One way to determine the sale price is to find 20% of $50 and then subtract that discount from $50. However, there's a faster, one-step way to solve this problem! If the printer ink is 20% off, that means that Sam is going to be paying 80% of the ink's original price. Therefore, if you calculate 80% of $50, you will have the sale price in just one step. 80% of

\$50 is $0.80 \times \$50 = \$40$, so the sale price is \$40. You can always use this method, by calculating the percentage that you *will* be paying, instead of focusing on finding the discount. So if something is discounted by 30%, you can just find 70% of the original cost to find the sale price. Use this shortcut to solve the next sample question.

Sample Question 5

You are buying a laser printer for your office. Printer option A is \$1,200 and 40% off and printer option B is \$950 and 20% off. What will the sale prices be of each printer? Which one will end up being less expensive?

Percent of Change (% Increase and % Decrease)

If I tell you that something is "\$10 off," can you say for certain whether this is a worthwhile discount? Since you don't know whether I'm offering \$10 off a \$12 meal or \$10 off a \$3,200 bike, hopefully your answer was "no!" Without knowing what the *original* cost was, it's not possible to judge whether a \$10 decrease in price is a worthy discount. The concept that we are now getting into is **percent of change**. Percent of change is a measure that compares the *amount of change* to the *original amount.*

Whether you are finding the **percent increase** or **percent decrease**, the percent of change is easiest to find by using the formula:

$$\% \text{ of change} = \frac{\text{amount of change}}{\text{original amount}} \times 100.$$

In this formula, the *amount of change* is the exact amount of decrease or increase, which is calculated by subtracting the original and new amount.

Example

Last week Gina's pizza shop sold an average of 10 pizzas during the lunch rush, but this week they sold an average of 8 pizzas during the same timeframe. By what percent did the pizza sales decrease from last week to this week?

1. Calculate the decrease (which is the *amount of change*): $10 - 8 = 2$
2. Put the *amount of change* and *original amount* into the percent change formula: $\% \text{ of change} = \frac{\text{amount of change}}{\text{original amount}} \times 100$

   $\% \text{ of change} = \frac{2}{10} \times 100$
3. Solve: $\% \text{ of change} = \frac{2}{10} \times \frac{100}{1} = \frac{200}{10} = 20$

So pizza sales decreased by 20%. (Remember that since you are already multiplying by 100, your answer is in decimal form and only needs the percent symbol to be added on.)

CAUTION

Doesn't it seem reasonable to assume that if a number decreases by 80% one week, and then increases by 80% the next week, that it will be back at the original number? Let's investigate this and see: Let's begin with a stock price of $100. An 80% drop in price would mean an $80 decrease (since 80% of $100 is $80), so the new stock price would be $20. If the $20 stock price goes up by 80% the following week, it would increase by $16 since 80% of $20 = 0.80 × 20 = $16. Therefore, an 80% increase would only raise the stock price to $36 instead of to the desired $100! So although it seemed reasonable that an 80% decrease followed by an 80% increase would bring a price back up to the same amount, this was a false assumption! This is false because the second week's increase of 80%, is 80% of a much *smaller number*, so the increase was not the same as the original 80% decrease.

Exploring what you just learned in the previous CAUTION box, let's see the percentage increase that would be necessary for Gina's lunch rush pizza sales to increase from 8 back up to 10:

1. **Calculate the increase needed (which is the *amount of change*):** $10 - 8 = 2$
2. **Put the *amount of change* and *original amount* into the percent change formula:** $\% \text{ of change} = \frac{\text{amount of change}}{\text{original amount}} \times 100$

   $\% \text{ of change} = \frac{2}{8} \times 100$

3. Solve: $\% \text{ of change} = \frac{2}{8} \times \frac{100}{1} = \frac{200}{8} = 25$

So in order for the pizza sales to return to their original average of 10 pizzas, the sales need to now increase by 25%.

Sample Question 6

Ryan bought a stock for $99 in March. It dropped to $66 in April. In May it went up to $88. Find the percent of decrease from March to April and the percent increase from April to May and compare.

Practice

Find the percent of change. If the percentage doesn't come out evenly, round to the nearest tenth of a percent.

_____ **12.** A population change from 8.2 million people to 7.4 million people

_____ **13.** From 15 miles per gallon to 27 miles per gallon

_____ **14.** From a police force of 120 officers to 150 officers

_____ **15.** A change from $2\frac{3}{4}$ cups of flour to $3\frac{1}{4}$ cups

_____ **16.** A change from $\frac{1}{2}$ of the group voting for Bill A to $\frac{3}{5}$ of the group voting for Bill A

_____ **17.** Barb tells Jan that she got a fantastic deal since she bought a $20 T-shirt that was on sale for $10. Jan claims her purchase of a $75 shirt that was on sale for $65 was just as good since she also saved $10. Calculate the percent of change in price for both shirts and compare.

_____ **18.** A new company, Ebibana, has an initial stock offering to employees for $8 a share and Lily spends half of her paycheck buying stock for herself. After a month of being available to the public, the stock plummets to $4 a share. Calculate the percent decrease from $8 to $4 and then determine the percent increase that is necessary for the stock price to return to the original price that Lily paid per share.

TIP

Here's a shortcut to finding the total price of an item with tax. Write the tax as a decimal, and then add it to 1 (8% tax would become 1.08). Then, multiply that by the cost of the item. The product is the total price. Example: What would a $34 vase cost if tax is 6.5%? $34 × 1.065 = $36.21, which would be the final cost. (This trick works, because the "1" is representing the base cost of the vase and the .065 is representing the tax on the vase, so when you multiply the vase by 1.065 your answer includes the vase *and* the tax.)

Percent Word Problems

Use the translation approach to solve these word problems. Before doing the translation, remember to boil the given information down to a single essential question.

_____ **19.** A \$180 suit is discounted 15%. What is the sale price?

_____ **20.** Ron started the day with \$150 in his wallet. He spent 9% of it to buy breakfast, 21% to buy lunch, and 30% to buy dinner. If he didn't spend any other money that day, how much money did he have left at the end of the day?

_____ **21.** Jacob invested \$20,000 in a new company that paid 10% interest per year on his investment. He did not withdraw the first year's interest, but allowed it to accumulate with his investment. However, after the second year, Jacob withdrew all his money (original investment plus accumulated interest). How much money did he withdraw in total?

_____ **22.** If Sue sleeps six hours every night, what percentage of her day is spent sleeping?

_____ **23.** Linda purchased \$500 worth of stocks on Monday. On Thursday, she sold her stocks for \$600. What was the percent increase on her investment?

_____ **24.** The Compuchip Corporation laid off 20% of its 5,000 employees last month. How many employees were NOT laid off?

_____ **25.** A certain credit card company charges $1\frac{1}{2}$% interest per month on the unpaid balance. If Joni has an unpaid balance of \$300, how much interest will she be charged for one month?

_____ **26.** Chris has a car payment that is $380 a month. If he is more than 5 days late with his payment, a $5% fine and a $25 late fee get added to his payment. If Chris makes his car payment 8 days late, how much will he owe?

_____ **27.** The cost of a home purchased in 2013 was $380,000. If it increases to $418,000 by 2015, what will be the percent increase?

_____ **28.** Lisa just sold an 18-carat gold bracelet for $464. If she bought it three years ago for $580, what was the percent decrease of value?

TRY THIS

The next time you eat in a restaurant, figure out how much of a tip to leave your server without using a calculator. In fact, figure out how much 15% of the bill is and how much 20% of the bill is, so you can decide how much tip to leave. Perhaps your server was a little better than average, so you want to leave a tip slightly higher than 15%, but not as much as 20%. If that's the case, figure out how much money you should leave as a tip. Do you remember the shortcut for figuring tips from this lesson?

Answers

Practice Problems

1. 150 students
2. 17.5%
3. 125 pens
4. $7.65 tax
5. 5%
6. $400
7. $3.75
8. $320
9. $17.28 or $17.30
10. $3,000
11. $0.92 or $0.90
12. 9.8% decrease
13. 80% increase
14. 25% increase
15. 18.2% increase
16. 20% increase
17. Barb saved 50% and Jan saved $13\frac{1}{3}$%, so Barb got a much better deal.
18. Percent change from $8 to $4: 50% decrease. Percent change needed to go from $4 to $8: 100% increase.
19. $153
20. $60
21. $24,200
22. 25%
23. 20% increase
24. 4,000
25. $4.50
26. $424
27. 10% increase
28. 20% decrease

Sample Question 1

Translate:

- ***90%*** is equivalent to both 0.9 and $\frac{9}{10}$
- ***of the 300 dentists*** means × *300*
- ***How many dentists*** is the unknown quantity: We'll use *d* for it.

But, wait! ***Ninety percent of the dentists did recommend sugarless gum***, but we're asked to find ***the number of dentists who did NOT recommend it***. So there will be an extra step along the way. You could determine how many dentists did recommend sugarless gum, and then subtract from the total number of dentists to find out how many did not. But there's an easier way:

Subtract 90% (the percent of dentists who did recommend sugarless gum) from 100% (the percent of dentists surveyed) to get 10% (the percent of dentists who did NOT recommend sugarless gum).

There's one more translation before you can continue: ***10%*** is equivalent to both 0.10 (which is the same as 0.1) and $\frac{10}{100}$ (which reduces to $\frac{1}{10}$). Now, put it all together and solve for *d*:

$0.1 \times 300 = d$ or $\frac{1}{10} \times \frac{300}{1} = d$

$30 = d$ $\quad$ $\frac{30}{1} = d$

Thus, **30** dentists did NOT recommend sugarless gum.

Sample Question 2

Although you have learned that ***of*** means *multiply*, there is an exception to the rule. The phrase ***out of*** means *divide*; specifically, ***2 out of 1,000 light bulbs*** means $\frac{2}{1,000}$ of the light bulbs are defective. We can equate (=) the fraction of the defective light bulbs ($\frac{2}{1,000}$) to the unknown percent that is defective, or $\frac{d}{100}$. (Remember, a percent is a number divided by 100.) The resulting equation and its solution are shown here.

Translate: $\frac{2}{1,000} = \frac{d}{100}$

Cross multiply: $2 \times 100 = 1,000 \times d$

$200 = 1,000 \times d$

Solve for d: $200 = 1,000 \times \mathbf{0.2}$

Thus, **0.2%** of the light bulbs are assumed to be defective.

Sample Question 3

Translate:

- Tax = $8\frac{1}{2}$%, which is equivalent to both $\frac{8\frac{1}{2}}{100}$ and 0.085
- Tax = $600
- Sales is the unknown amount; we'll use S to represent it.
- Tax = $8\frac{1}{2}$% ***of*** sales ($\times S$).

Fraction approach:

Translate: $600 = \frac{8\frac{1}{2}}{100} \times S$

Rewrite 600 and S as fractions: $\frac{600}{1} = \frac{8\frac{1}{2}}{100} \times \frac{S}{1}$

Multiply fractions: $\frac{600}{1} = \frac{8\frac{1}{2} \times S}{100}$

Cross multiply: $600 \times 100 = 1 \times 8\frac{1}{2} \times S$

Solve for S by dividing both sides of the equation by $8\frac{1}{2}$:

$60{,}000 = 8\frac{1}{2} \times S$

$60{,}000 \div 8\frac{1}{2} = 8\frac{1}{2} \times S \div 8\frac{1}{2}$

$7{,}058.82 \approx S$

Decimal approach:

Translate and solve for S by dividing by 0.085:

$600 = 0.085 \times S$

$600 \div 0.085 = 0.085 \times S \div 0.085$

Rounded to the nearest cent and excluding tax,

$7{,}058.82 \approx S$

$7,058.82 is the amount of sales on May 1.

Sample Question 4

- First round 12,013 voters to 12,000.
- 10% of 1,200 is 1,200, so 5% is 600.
- 65% = (10% × 6) + 5%
- So 65% of 12,000 = (1,200 × 6) + 600 = 7,800
- Therefore, approximately 7,800 people turned up to vote in Hamilton County

Sample Question 5

- Printer option A is 40% off of $1,200, so you will pay 60% of $1,200:
- 60% of $1,200 = 0.60 × $1,200 = $720 for Printer A
- Printer option B is 20% off of $950, so you will pay 80% of $950
- 80% of $950 = 0.80 × $950 = $760 for Printer B
- So although Printer A is originally more expensive, with the current discounts being offered it will be less expensive than Printer B.

Sample Question 6

- Calculate the decrease (which is the *amount of change*): $\$99 - \$66 = \$33$
- Put the *amount of change* and *original amount* into the percent change formula: $\%\text{ of change} = \frac{\text{amount of change}}{\text{original amount}} \times 100$

  $\%\text{ of change} = \frac{33}{99} \times 100$
- Solve: $\%\text{ of change} = \frac{33}{99} \times \frac{100}{1} = \frac{1}{3} \times \frac{100}{1} = \frac{100}{3} = 33\frac{1}{3}\%$

So the stock decreased by $33\frac{1}{3}\%$ from March to April.

Next, calculate the percent increase from \$66 to \$88:

- Calculate the increase (which is the *amount of change*): $\$88 - \$66 = \$22$
- Put the *amount of change* and *original amount* into the percent change formula: $\%\text{ of change} = \frac{\text{amount of change}}{\text{original amount}} \times 100$

  $\%\text{ of change} = \frac{22}{66} \times 100$
- Solve: $\%\text{ of change} = \frac{22}{66} \times \frac{100}{1} = \frac{1}{3} \times \frac{100}{1} = \frac{100}{3} = 33\frac{1}{3}\%$

So the stock increased by $33\frac{1}{3}\%$ from April to May. Although the stock decreased by $33\frac{1}{3}\%$ and then increased by the same percentage, it is still not at its original value of \$99.

CHAPTER

12 Ratios and Proportions

Mathematics is a language.

—Josiah Willard Gibbs, theoretical physicist (1839–1903)

CHAPTER SUMMARY

Your mind thinks in terms of ratios all the time and you probably don't even realize it. This chapter explores ratios, and how they are used in the office and at home. It also demonstrates how ratios are the building blocks of proportions, which are another necessary tool to real-world problem solving.

Ratios

A ratio is a way of comparing two numbers, and showing how the quantities of different groups compare to one another. A ratio can compare the *part* of something to the *whole*, but it is more common for a ratio to compare two different *parts* of the same thing. Look at these examples:

Part to Whole: *There are 24 tablespoons of sugar in every 1 gallon of Puckerface Lemonade.* (comparing 24 to 1)
Part to Part: *Puckerface Lemonade has 24 tablespoons of sugar for every 32 tablespoons of lemon juice.* (comparing 24 to 32)

Ratios can even be used to compare three different *parts*:

Part to Part to Part: *Puckerface Lemonade Recipe: For every 16 cups of water, use 24 tablespoons sugar and 32 tablespoons lemon juice.* (comparing 16 to 24 to 32)

We will come back to the Puckerface Lemonade recipe later, but for now, let's have a look at the Northport Bocce Ball Club, which states on its website, "*Our male to female ratio is 3 to 5.*" Although at first it might seem like there are just 8 people in the club, this statement doesn't mean that there are *only* 3 men and 5 women. What this means is that for *every* grouping of 3 men, there is a grouping of 5 women. Therefore, if there were 6 men in the club, there would be 10 women (since this would be two groups of 3 men and two groups of 5 women). The following table illustrates some of the different possibilities of groups of men and women that could comprise the Northport Bocce Ball Club.

BREAKDOWN OF CLUB MEMBERS; 3 TO 5 RATIO

# OF GROUPS	# OF MEN	# OF WOMEN	TOTAL MEMBERS
1	mmm	wwwww	8
2	mmm mmm	wwwww wwwww	16
3	mmm mmm mmm	wwwww wwwww wwwww	24
4	mmm mmm mmm mmm	wwwww wwwww wwwww wwwww	32
5	mmm mmm mmm mmm mmm	wwwww wwwww wwwww wwwww wwwww	40

As you look at the table, do you see a pattern in the number of total members? When there is just a single group of 3 men and 5 women, there are 8 members. When there are two groups of 3 men and 5 women, there are 16 members. Each time there is an extra group of 3 men and 5 women, the total number of members increases by 8.

The ratio "*3 men for every 5 women*" can be expressed in several different ways:

- By using the word "**to**": *3* ***to*** *5*
- By using a colon: 3 : 5
- By using a fraction: $\frac{3}{5}$
- By using "**out of**" (when comparing a *part* to the *whole*): *3* ***out of*** *8*

Looking at the preceding table, notice that the first row has a ratio of 3 : 5 or $\frac{3}{5}$ men to women. Looking at the second row, the ratio can be written as 6 : 10 or $\frac{6}{10}$. Remembering what was presented about reducing fractions to lowest terms, you should be able to recognize that $\frac{6}{10}$ is equivalent to $\frac{3}{5}$. Therefore, regardless of whether there are 6 men and 10 women, 9 men and 15 women, or 12 men and 20 women, the ratio of men to women is always going to reduce down to 3 : 5. Like fractions, ratios should always be written in lowest terms.

Now that you know that ratios should always be in lowest terms, let's go back to the Puckerface Lemonade recipe where a gallon of lemonade requires 24 tablespoons sugar and 32 tablespoons lemon juice. Rather than referring to the sugar to lemon juice ratio as 24 : 32, it should be expressed as 3 : 4. (24 : 32 is 8 groups of 3 : 4).

CAUTION

The order in which you write the numbers in a ratio is critical to its meaning. Saying that the "*ratio of men to women is 3 to 5*" is much different than saying it is "*5 to 3.*" Regardless of whether you are writing the ratio as a fraction, with a colon, or with the word "*to*," you must be careful that the order in which you write the numbers clearly matches the order in which the descriptor words are being used to identify each group.

Here are some examples of how ratios are used to model everyday situations. Notice how given data can be boiled down to a reduced fraction ratio, and how, conversely, a ratio can be raised to higher terms to model larger scale numbers.

- Last year, it snowed 13 out of 52 weekends in New York City. The ratio *13 out of 52* can be reduced to lowest terms (*1 out of 4*) and expressed as any of the following:

  1 to 4
1:4
$\frac{1}{4}$ } Reducing to lowest terms tells you that it snowed 1 out of 4 or $\frac{1}{4}$ of the weekends.

- Lloyd drove 140 miles on 3.5 gallons of gas, for a ratio (or gas *consumption rate*) of 40 *miles per gallon*:

  $$\frac{\overset{40}{\cancel{140}}\ \textit{miles}}{\underset{1}{\cancel{3.5}}\ \textit{gallons}} = \frac{40\ \textit{miles}}{1\ \textit{gallon}} = 40\ \textit{miles per gallon}$$

- The teacher-student ratio at Bronxville Preschool is 1 to 7, which means there are 7 students for every 1 teacher. For example, if the school has 20 teachers, there are 140 students. (This would mean that there are 20 groups, each having 1 teacher and 7 students.)

- Pearl's Pub has 5 chairs for every table. If it has 100 chairs, then it has 20 tables.
- Jorge used 20 large bags of Del Rio pebbles to cover a fire pit enclosure that is 30 square feet. Therefore the ratio of pebbles to square feet used was 20 : 30, or 2 : 3.

In the previous lesson you learned that the word "is" can be translated into "=" and the word "of" means multiplication. Another translation is the word *per*, which means that two numbers can be expressed as a ratio, normally with the number 1 as the denominator. For example, *Jeff averaged 80 miles per hour,* can be expressed as $\frac{\text{80 miles}}{\text{1 hour}}$. Here are some other examples:

$$24 \text{ miles per gallon} = \frac{24 \textit{ miles}}{1 \textit{ gallon}}$$

$$\$12 \text{ per hour} = \frac{12 \textit{ dollars}}{1 \textit{ hour}}$$

$$3 \text{ meals per day} = \frac{3 \textit{ meals}}{1 \textit{ day}}$$

$$4 \text{ cups per quart} = \frac{4 \textit{ cups}}{1 \textit{ quart}}$$

TIP

As we saw in the Puckerface Lemonade recipe, ratios can also be used to relate more than two items. "*Puckerface Lemonade Recipe: For every 16 cups of water, use 24 tablespoons sugar and 32 tablespoons lemon juice.*" When expressing 3-part ratios, fractions cannot be used, so use either a colon or the world "to," and don't forget to reduce the ratio to lowest terms. Therefore since Puckerface Lemonade has a water to sugar to lemon juice ratio of 16 : 24 : 32, that would be correctly reduced to 2 : 3 : 4, by dividing all the quantities by 8.

Practice

Write each of the following as a ratio.

_____ **1.** A cleaning solution uses 2 parts of concentrated soap to 5 parts water.

_____ **2.** When ordering desserts for a banquet you need 6 cookies for every 3 cupcakes and 2 brownies.

_____ **3.** 56 people must fit into 12 cars.

_____ **4.** Nurses at Singing Cedar Eldercare work 3 double shifts every 7 days.

_____ **5.** During a conference Cole's Coffee Cart used 12 cups of sugar for the 200 cups of coffee sold.

_____ **6.** In a trial of 30 coin flips, the referee got 15 heads and 15 tails.

_____ **7.** The Santa Monica street parking costs $0.25 for every 8 minutes. (It is not acceptable to write ratios with decimals, so make sure to convert this into whole numbers)

_____ **8.** A slow-cooker recipe calls for a cooking time of 40 minutes for every $\frac{1}{2}$ pound of meat. (It is not acceptable to use a fraction *within* a ratio, so make sure to convert this into whole numbers)

Finish the comparison.

_____ **9.** If 3 out of 5 people pass this test, how many people will pass the test when 45 people take it?

_____ **10.** The ratio of professors to students at a private university is 1 : 13. If there are 3,600 students enrolled at the university, how many professors work there?

Ratios and Totals

A ratio usually tells you something about the total number of things being compared. In our first ratio example of the Northport Bocce Ball Club with 3 men for every 5 women, the club's total membership is a **multiple of 8** because each group contains 3 men and 5 women. When the two parts in a ratio make up the whole (such as "men" and "women" in a club), you can divide the total number by the sum of the parts in the ratio to see how many subgroups make up the whole. For example, if you found out that the Northport Bocce Ball Club had 80 members, you could divide 80 by the 8 parts that make up the ratio 3 : 5. This will reveal that there are 10 subgroups of 3 men and 5 women. From there you could conclude that there must be $3 \times 10 = 30$ men and $5 \times 10 = 50$ women in the club. The following example shows the steps to solving *total* questions.

Example

Wyatt ordered 245 new books for the library this year. For every 3 fiction books he ordered, he also ordered 2 nonfiction books. How many fiction and nonfiction books did Wyatt order?

1. First determine how many items each subgroup contains: Each subgroup contains 5 books since there are 3 fiction books and 2 nonfiction books.
2. Divide the total number of books (245) by the number of books in each subgroup (5) to determine to determine how many subgroups make up the whole: $245 \div 5 = 49$ subgroups.

3. Multiply the number of subgroups (49) by the number of fiction books in each subgroup (3) to determine the number of fiction books: $49 \times 3 = 147$ fiction books.
4. Multiply the number of subgroups (49) by the number of nonfiction books in each subgroup (2) to determine the number of nonfiction books: $49 \times 2 = 98$ nonfiction books.
5. Check your answer by making sure that both groups add up to the total: $147 + 98 = 245$.

Therefore, Wyatt ordered 147 fiction books and 98 nonfiction books.

When a ratio involves three different things, the sum of the three becomes your multiple. For example, if there are 72 cars and the ratio of red to blue to white cars is 5:1:3, then the cars can be broken into multiples of 9.

Now you try working with ratios and totals with the following sample question. The step-by-step solution is at the end of the lesson.

Sample Question 1

Every day, Bob's Bakery makes fresh cakes, pies, and muffins in the ratio of 3:2:5. If a total of 300 cakes, pies, and muffins is baked on Tuesdays, how many of each item is baked?

Practice

Try these ratio word problems.

_____ **11.** Agatha died and left her $40,000 estate to her friends Bruce, Caroline, and Dennis in the ratio of 13:6:1, respectively. How much is each friend's share? (Hint: The word respectively means that Bruce's share is 13 parts of the estate because he and the number 13 are listed first, Caroline's share is 6 parts because both are listed second, and Dennis's share is 1 part because both are listed last.)

_____ **12.** There were 28 people at last week's board meeting. If the ratio of men to women was 4:3, how many women were at the meeting?

_____ **13.** At a certain corporation, the ratio of clerical workers to executives is 7 to 2. If a combined total of 81 clerical workers and executives work for that corporation, how many clerical workers are there?

_____ **14.** Kate invests her retirement money with certificates of deposit, low-risk bonds, and high-risk stocks in a ratio of 3:5:2. If she put aside $8,000 last year, how much was invested in high-risk stocks?

_____ **15.** A unit price is a ratio that compares the price of an item to one single unit of measurement. Calculate the unit price of each item by determining how much *one ounce* costs in each package with division. Which of these five boxes of Klean-O Detergent is the best buy?

a. Travel-size: $1 for 5 ounces
b. Small: $2 for 11 ounces
c. Regular: $4 for 22 ounces
d. Large: $7 for 40 ounces
e. Jumbo: $19 for 100 ounces

_____ **16.** Shezzy's pulse rate is 19 beats every 15 seconds. What is his rate in beats per minute?

Proportions

A *proportion* is an equation in which two ratios are set equal to each other. Proportions are commonly used to solve real-world problems when a fixed ratio is used to calculate a ratio in greater quantities. You probably think in terms of proportions on occasion without realizing it. If a recipe makes 12 cupcakes, but you want to make 36 cupcakes, knowing that you need to multiply all the ingredients by 3 is actually a type of proportion. Let's look at the following statement and how proportions could be applied to it:

Nine out of ten professional athletes suffer at least one injury each season.

The phrase *nine out of ten* is a ratio. It tells you that $\frac{9}{10}$ of professional athletes suffer at least one injury each season. But there are more than 10 professional athletes. Suppose that we want to apply this ratio to 120 athletes to determine how many of them will be likely to have an injury in the upcoming season. A proportion can be set up that puts the number of *injured* athletes in the numerator and the *total* number of athletes in the denominator:

$$\frac{9 \text{ injured}}{10 \text{ total}} = \frac{? \text{ injured}}{120 \text{ total}}$$

Hopefully you remember that you can use equivalent fractions here to determine what the new numerator would be. In this case, multiply the numerator and denominator by 12 to create the equivalent fraction:

$$\frac{9 \text{ injured}}{10 \text{ total}} \times \frac{12}{12} = \frac{108 \text{ injured}}{120 \text{ total}}$$

Therefore, out of 120 professional athletes, it is likely that 108 of them will suffer from an injury in any given season.

Sometimes it is not so easy to recognize how to create the equivalent fraction to a given ratio, which is why the helpful technique presented next is an important skill to master.

Cross Products

Many proportion word problems are easily solved with fractions and cross products. As you probably recall from the lessons on fractions, the cross products of proportions are equal:

$$\frac{9 \text{ injured}}{10 \text{ total}} \times \frac{108 \text{ injured}}{120 \text{ total}}$$

$$9 \times 120 = 1{,}080$$

$$10 \times 108 = 1{,}080$$

When setting up a proportion, it very important that both ratios are written with like items in the same respective place.

For example, let's say we have two ratios (ratio #1 and ratio #2) that compare red marbles to white marbles. When you set up the proportion, both fractions must be set up the same way—with the red marbles on top and the corresponding white marbles on bottom, or with the white marbles on top and the corresponding red marbles on bottom:

$$\frac{red_{\#1}}{white_{\#1}} = \frac{red_{\#2}}{white_{\#2}} \text{ or } \frac{white_{\#1}}{red_{\#1}} = \frac{white_{\#2}}{red_{\#2}}$$

Alternatively, one fraction may compare the red marbles while the other fraction compares the white marbles, with both comparisons in the same order:

$$\frac{red_{\#1}}{red_{\#2}} = \frac{white_{\#1}}{white_{\#2}} \text{ or } \frac{red_{\#2}}{red_{\#1}} = \frac{white_{\#2}}{red_{\#2}}$$

Let's revisit the Puckerface Lemonade recipe and see how to set up and solve a proportion.

Example

Puckerface Lemonade calls for a sugar to lemon juice ratio of 3 to 4. Heidi wants to make as much lemonade as possible but she only has 50 tablespoons of lemon juice. How much sugar will she need for her lemonade?

Solution

1. Set up a proportion where the ratio of the recipe equals the ratio of Heidi's ingredients. The ratio representing the recipe will have the sugar in the numerator and the lemon juice in the denominator. This means that the ratio representing Heidi's ingredients must also have the sugar and lemon juice in the corresponding positions:

$$\frac{3 \text{ T sugar}}{4 \text{ T lemon juice}} = \frac{? \text{ T sugar}}{50 \text{ T lemon juice}}$$

2. Use cross products to set up a new equation:

$$\frac{3}{4} = \frac{s}{50}$$

$$3 \times 50 = 4s$$

$$150 = 4s$$

3. Solve the equation by dividing both sides by the number that is being multiplied to the variable:

$$150 = 4s$$

$$\underline{\div 4 \;\; \div 4}$$

$$37.5 = s$$

Therefore, Heidi must use 37.5 tablespoons of sugar with her 50 tablespoons of lemon juice.

Sample Question 2

The ratio of men to women at a certain meeting is 3 to 5. If there are 18 men at the meeting, how many people are at the meeting?

Practice

Use cross products to find the unknown quantity in each proportion.

_____ **17.** $\frac{120 \text{ miles}}{2.5 \text{ gallons}} = \frac{m \text{ miles}}{1 \text{ gallon}}$

_____ **18.** $\frac{135 \text{ miles}}{h \text{ hours}} = \frac{45 \text{ miles}}{1 \text{ hour}}$

_____ **19.** $\frac{150 \text{ drops}}{t \text{ teaspoon}} = \frac{60 \text{ drops}}{4 \text{ teaspoons}}$

_____ **20.** $\frac{20 \text{ minutes}}{\frac{1}{4} \text{ pound}} = \frac{m \text{ minutes}}{2\frac{1}{2} \text{ pounds}}$

Try these proportion word problems.

_____ **21.** A certain model of calculator is known to have approximately 7 malfunctioning calculators out of every 2,000. If an office supply store in Los Angeles is going to order 9,000 of these calculators to stock their shelves for back-to-school shopping in August, approximately how many returns due to defects should the store anticipate having later in the fall?

_____ **22.** In an enlarged photograph of the New York City skyline, every centimeter represents $18\frac{3}{4}$ feet. If the Chrysler Building is 56 cm tall in this photo, approximately how tall is it in real life?

_____ **23.** The last time Elizabeth had friends over, she noticed that they ate 2 bowls of mixed nuts for every 5 bowls of popcorn. If all together they had 21 bowls of snack food, and tomorrow she is having twice the number of guests as last time, how many bowls of mixed nuts should she provide?

_____ **24.** The Robb family wants to have the carpet in their vacation home steam cleaned. They received a special offer in the mail advertising $3 for every 10 square feet of carpeting cleaned. If they don't want to spend more than $250, how many square feet of carpeting can they have cleaned with this offer?

_____ **25.** A marinade recipe calls for $1\frac{1}{3}$ cups of soy sauce for every $\frac{1}{2}$ cup of water and $\frac{3}{4}$ cup of sesame oil. If this recipe is enough to marinate 16 ounces of tofu, but Paul wants to only marinate 12 ounces of tofu, how much soy sauce should he use?

_____ **26.** The Texas State lottery advertises that for every 4 losing tickets, there are 3 winning tickets where prize winners earn $0.50 or more. If they sold 504 lottery tickets last week, how many winners were there?

TIP

Go to a grocery store and look closely at the prices listed on the shelves. Pick out a type of food that you would like to buy, such as cold cereal, pickles, or ice cream. To determine which brand has the cheapest price, you need to figure out each item's unit price. The unit price is a ratio that gives you the price per unit of measurement for an item. Without looking at the tags that give you this figure, calculate the unit price for three products, using the price and size of each item. Then, check your answers by looking at each item's price label that specifies its unit price.

Answers

Practice Problems

1. $2:5$ or $\frac{2}{5}$
2. $3:2:1$
3. $14:3$ or $\frac{14}{3}$
4. $3:7$ or $\frac{3}{7}$
5. $3:50$ or $\frac{3}{50}$
6. $1:1$ or $\frac{1}{1}$
7. $1:32$ or $\frac{1}{32}$
8. $80:1$ or $\frac{80}{1}$
9. 27 will pass
10. 277 professors
11. Bruce: $26,000, Caroline: $12,000, Dennis: $2,000
12. 12 women
13. 63 clerical workers
14. $1,600 in high-risk stocks
15. Large is the best deal at 17.5 cents per ounce.
16. 76 beats per minute
17. 48 miles
18. 3 hours
19. 10 teaspoons
20. 200 minutes
21. Since you cannot have 31.5 calculators, the store should expect 31 or 32 returns.
22. 1,050 feet
23. 12 bowls of mixed nuts
24. 833 square feet of carpet
25. 1 cup
26. 216 winners

Sample Question 1

1. First determine how many items each subgroup contains:
   - 3 + 2 + 5 = 10 baked items
2. Divide the total number of baked items (300) by the number of items in each subgroup (10) to determine how many subgroups make up the whole:
   - 300 ÷ 10 = 30 subgroups
3. Multiply the number of subgroups (30) by the number of cakes (3), pies (2), and muffins (5) in each subgroup to determine how many of each type of baked goods were in the whole:
   - 30 × 3 = 90 cakes
   - 30 × 2 = 60 pies
   - 30 × 5 = 150 muffins
4. Check your answer by making sure that all the groups add up to the total:
   - 90 + 60 + 150 = 300

Sample Question 2

The first step of the solution is finding a fraction equivalent to $\frac{3}{5}$ with 18 as its top number (because both top numbers must reflect the same thing—in this case, the number of men). Since we don't know the number of women at the meeting, we'll use the unknown *w* to represent them. Here's the mathematical setup and solution:

$$\frac{3\text{ men}}{5\text{ women}} = \frac{18\text{ men}}{w\text{ women}}$$

$$\frac{3}{5} = \frac{18}{w}$$

$$3 \times w = 5 \times 18$$

$$3 \times w = 90$$

$$3 \times 30 = 90$$

Since there are 30 women and 18 men, a total of 48 people are at the meeting.

Check:

Reduce $\frac{18}{30}$. Since you get $\frac{3}{5}$ (the original ratio), the answer is correct.

CHAPTER

13 Averages: Mean, Median, and Mode

Poor teaching leads to the inevitable idea that the subject (mathematics) is only adapted to peculiar minds, when it is the one universal science and the one whose . . . ground rules are taught us almost in infancy and reappear in the motions of the universe.

—Henry J. S. Smith, Irish mathematician (1826–1883)

CHAPTER SUMMARY

This chapter focuses on three key measures that are often used to describe data. These numbers are sometimes referred to as *measures of central tendency* and are different types of averages. This lesson defines mean, median, and mode; explains the differences among them; and shows you how to use them.

An *average* is a number that *typifies* or represents a group of numbers. Averages are used in the workplace on a regular basis—the average number of pizzas sold during a lunch hour, the average number of toys manufactured per week in a factory, the average starting salary for your level of schooling, the average number of miles per gallon of gas a car gets, and so forth.

There are three different measures of central tendency that are most commonly used to summarize a group of numbers:

1. the ***mean***
2. the ***median***
3. the ***mode***

Although all these can be considered a type of *average*, when you hear people mention an average, the majority of the time they are referring to the ***mean*** and not to the other two measures of central tendency. Therefore, whenever we use the word *average* in this book, we are referring to the *mean*.

In order to compare and contrast the three different measures of central tendency, we will use the same group of numbers to compute the ***mean***, ***median***, and ***mode***. The following table shows how many students are in 9 different classrooms at the Chancellor School. Since the mean is the most commonly used measure, we will begin there.

ROOM #	1	2	3	4	5	6	7	8	9
# of students	15	15	11	16	15	17	16	30	18

Mean (Average)

You probably have heard the word *average* lots of times and it's also likely that you have some experience calculating the average of a group of numbers. The ***mean*** of a group of numbers is the *arithmetic average*, and it is calculated by dividing the *sum* of the numbers by the *number* of numbers:

$$\text{Average} = \frac{\text{Sum of the numbers}}{\text{Number of numbers}}$$

Example

Find the **average** number of students in a classroom at the Chancellor School.

Solution

$$\text{Average} = \frac{15 + 15 + 11 + 16 + 15 + 17 + 16 + 30 + 18}{9} = \frac{153}{9}$$

Average = 17

The **average** (**mean**) number of students in a classroom at the Chancellor School is **17.** Do you find it curious that only two classrooms have more students than the average or that the average isn't right smack in the middle of the list of numbers being averaged? Read on to find out about a measure that is right in the middle of things.

Median

The median of a group of numbers is the number in the middle when the numbers are arranged in order.

When there is an odd number of data, the median will be the center number that has an equal number of data on either side of it. When there is an even number of data, the median is the average of the two middle numbers.

Example

Find the **median** number of students in a classroom at the Chancellor School.

Solution

Simply list the numbers in order (from low to high) and identify the number in the middle:

11 15 15 15 $\boxed{16}$ 16 17 18 30

In the case of the Chancellor School, the ***median*** number of students per classroom is 16.

Notice that if there had been an even number of classrooms, then there would have been two middle numbers.

Here we will add a tenth classroom that has 9 students in it to demonstrate how that would change the median:

$$\mathbf{15\frac{1}{2}}$$

$$\downarrow$$

$$9 \quad 11 \quad 15 \quad 15 \quad \boxed{15} \quad \boxed{16} \quad 16 \quad 17 \quad 18 \quad 30$$

With ten classrooms instead of nine, the median is the average of 15 and 16, or $15\frac{1}{2}$, which is also halfway between the two middle numbers.

CAUTION

You must arrange the numbers in order when computing a median, but not a mean or mode.

Now let's look at *why* the median is a significant measure of central tendency. In order to do so we must first discuss *outliers*, which are pieces of data that are uncharacteristically above or below the majority of the data points. When interpreting data, *outliers* are significant because they *skew*, or pull, the *mean* (average) of the data up or down. For example, let's consider a scenario where Gina sells anywhere from 11 to 13 pizzas every lunch hour for the first 5 days of the week, and then sells zero pizzas on the 6th and 7th days because the shop is closed for construction. In this case, the *average* lunch hour sales for the week would get pulled down significantly by the two days of zero sales. However, the construction would not significantly impact Gina's ***median*** pizza sales. Therefore, in this case, the question, "What is the average number of pizzas sold during lunch hour by Gina?" is best answered with the ***median*** and not the ***average*** or ***mean***.

Hopefully it is starting to become clear that the ***median*** is a useful measure of central tendency that can be more representative of data sets that have outliers. Here is another illustration that should help you see how the *mean* and *median* tell us different things about data sets.

Consider the annual income of the residents of a major metropolitan area. A few multimillionaires could substantially raise the *average* annual income, but they would have no more impact on the *median* annual income than a few above-average wage earners. Thus, the *median* annual income is more representative of the residents than the *mean* annual income. In fact, you can conclude that the annual income for *half* the residents is greater than or equal to the median, while the annual income for the other half is less than or equal to the *median*. The same cannot be said for the *average* annual income.

TIP

Can you now see why the median is the most common type of average used to measure the price of homes in the real estate market? The median is a helpful tool because it is protected against skewed data that is very far from the true center number. (Like a neighborhood where most of the houses cost around $200,000, but there's one house for $750,000 because it has a horse stable.)

Mode

The mode of a group of numbers is the number(s) that appears most often.

Example
Find the **mode**, the most common classroom size, at the Chancellor School.

Solution
Scanning the data reveals that there are more classrooms with 15 students than any other size,
making **15** the mode:

11 [15] [15] [15] 16 16 17 18 30

Had there also been three classrooms of, say, 16 students, the data would be **bimodal**—both 15 and **16** are the modes for this group:

11 [15] [15] [15] [16] [16] [16] 17 18 30

However, had there been an equal number of classrooms of each size, the group would NOT have a mode—no classroom size appears more frequently than any other:

11 11 13 13 15 15 17 17 19 19

Hook
Here's an easy way to remember the definitions of median and mode.

Median: Picture a divided highway with a median running right down the **middle** of the road.

Mode: The **MO**de is the **MO**st popular member of the group.

Try your hand at this measures of central tendency question. Answers to sample questions are at the end of the lesson.

Sample Question 1

Dani is taking Organic Chemistry as part of the required courses to become a psychiatrist. Her test scores for the semester were: 95, 72, 94, 72, 91, 88, 94, 20, 93. Find the three measures of central tendency, mean, median, and mode. Which of these do you feel is the most representative number to summarize her success in the class?

TIP

Use these tricks to remember what each term means:

Mean: That mean teacher gave me a failing average!

Median: "Median" sounds like medium, which is the size in the middle of small, medium, large. So "median" is the number in the middle.

Mode: "Mode" sounds the closest to "most," so the mode is the number(s) that occurs the most.

Practice

This sales chart shows February's new car and truck sales for the top five sales associates at Vero Beach Motors:

CAR AND TRUCK SALES

	ARNIE	BOB	CALEB	DEBBIE	ED	WEEKLY TOTAL
Week 1	7	5	0	8	7	27
Week 2	4	4	9	5	4	26
Week 3	6	8	8	8	6	36
Week 4	5	9	7	6	8	35

_____ **1.** The monthly sales award is given to the sales associate with the highest weekly average for the month. Who won the award in February?

_____ **2.** What was the average of all the weekly totals?

_____ **3.** What was the median of all the weekly totals?

_____ **4.** What was the mean number of February car sales for each associate?

_____ **5.** What was the median number of February car sales for each associate?

_____ **6.** Based on the February sales figures, what was the most likely number of cars sold per week by any sales associate?

_____ **7.** Based on your previous answers, which measure of center do you think is most representative of the quality of the salesperson? Which person do you think may be the best salesperson and why?

Average Shortcut

If there's an *even-spacing* pattern in the group of numbers being averaged, you can determine the average without doing any arithmetic! For example, the following group of numbers has an *even-spacing* "3" pattern: Each number is 3 greater than the previous number:

6 9 12 15 **18** 21 24 27 30

The average is **18**, the number in the middle. When there is an even number of evenly spaced numbers in the group, there are two middle numbers, and the average is halfway between them:

6 9 12 15 **18** **21** 24 27 30 33

↑

$\mathbf{19\frac{1}{2}}$

This shortcut works even if each number appears more than once in the group, as long as each number appears the same number of times, for example:

10 10 10 20 20 20 30 **30** 30 40 40 40 50 50 50

Weighted Average

In a weighted average, some or all of the numbers to be averaged have a *weight* associated with them, meaning that some numbers count more than others because they have more significance or have occurred more often. Let's consider the following problem to see what weighted averages look like:

Example

Don averaged 50 miles per hour for the first three hours of his drive to Seattle. When it started raining, his average fell to 40 miles per hour for the next two hours. What was his average speed?

You cannot simply compute the average of the two speeds as $\frac{50 + 40}{2} = 45$, because Don spent more time driving 50 mph than he did driving 40 mph. In fact, Don's average speed is closer to 50 mph than it is to 40 mph precisely because he spent more time driving 50 mph. To correctly calculate Don's average speed, we have to take into consideration the number of hours at each speed: 3 hours at an average of 50 mph and 2 hours at an average of 40 mph, for a total of 5 hours.

$$\textit{Average} = \frac{50 + 50 + 50 + 40 + 40}{5} = \frac{230}{5} = 46$$

Or, take advantage of the weights, 3 hours at 50 mph and 2 hours at 40 mph:

$$\textit{Average} = \frac{(3 \times 50) + (2 \times 40)}{5} = \frac{230}{5} = 46$$

The second technique shown is particularly useful with larger sets of data, where it would be inconvenient to individually add up all the reoccurring numbers. In the following sample question, consider multiplying the number of students in the left column by the test scores in the right column in order to calculate the weighted average.

Sample Question 2

Find the average test score, the median test score, and the mode of the test scores for the 30 students represented in the following table.

Number of Students	Test Score
1	100
3	95
6	90
8	85
5	80
4	75
2	70
1	0

Practice

Use what you know about mean, median, and mode to solve these word problems.

_____ **8.** A university professor determines her students' grade by averaging the following three scores: the midterm, the final exam, and the student's average on four unit tests. Clive's average on the four unit tests is a 72. He scored a 53 on the midterm. What is the lowest score he can get on the final exam if he needs to pass the class with an average of 65?

_____ **9.** Luke times his first mile every time he goes running. Today he ran 4 miles, and his first mile was 6 minutes and 54 seconds. If his total time was 27 minutes and 12 seconds, what was his average mile pace on his last 3 miles?

_____ **10.** Find the mean and median of:

$\frac{1}{2}, \frac{1}{3}, \frac{1}{4}, \frac{1}{5}$

_____ **11.** What is the mean of: 0.03, 0.003, 0.3, 5.222, 0?

For Questions 12 through 14, use the following:

Tanya is baking 4 types of holiday cookies. The number of batches of each type of cookie and the amount of white sugar per batch are listed in the following:

TYPE	NUMBER OF BATCHES	AMOUNT OF SUGAR PER BATCH
Snickerdoodle	2	1 cup
Chocolate Chip	3	$1\frac{1}{3}$ cups
Oatmeal Raisin	1	$\frac{3}{4}$ cup
Peanut Butter	4	$\frac{2}{3}$ cup

_____ **12.** What is the average amount of sugar used in a batch of cookies?

_____ **13.** What is the median amount of sugar used in a batch of cookies?

_____ **14.** Suppose she increased the number of batches of each type of cookie by 1 to account for a larger group of people to whom she will distribute them. What is the new median amount of sugar?

Use the following table to answer questions 15 through 18.

PAGE COUNTY HOME SALES	
ZIP CODE	SALE PRICE
48759	$210,000
48577	$189,000
48383	$375,000
48759	$215,000
48750	$132,000
48759	$236,000
48577	$196,500
48577	$192,000

_____ **15.** Find the median house price in the table.

_____ **16.** Find the mean house price in the table, to the nearest dollar.

_____ **17.** By how much money does the mean home price drop when the mean is calculated after leaving out the most expensive home ($375,000)?

_____ **18.** Given the following group of numbers—8, 2, 9, 4, 2, 7, 8, 0, 4, 1—which of the following is (are) true?

I. The mean is 5.
II. The median is 4.
III. The sum of the modes is 14.

TRY THIS

Write down your age on a piece of paper. Next to that number, write down the ages of at least five of your family members. Find the mean, median, and mode of the ages you've written down. Remember that some groups of numbers do not have a mode. Does your group have a mode? If there are any outliers in your data, notice how they skew the mean, but not the median and mode.

Answers

Practice Problems

1. Debbie
2. 31
3. 31
4. Arnie: 5.5; Bob: 6.5; Caleb: 6; Debbie: 6.75; Ed: 6.25
5. Arnie: 5.5; Bob: 6.5; Caleb: 7.5; Debbie: 7; Ed: 6.5
6. There were 5 occasions where a sales associate sold 8 cars in a week, so the mode is 8.
7. Answers may vary. Since the mean can get pulled up by a lucky week or pulled down during a week of vacation, it might not the best indicator of sales performance and the median may be more reliable since that takes the average of the middle two weeks. It looks like Caleb and Debbie might be the best salespeople.
8. 70
9. 6 minutes 46 seconds
10. mean $\frac{77}{240}$; median $\frac{7}{24}$
11. 1.111
12. $\frac{15}{16}$ cups
13. $\frac{7}{8}$ cup
14. $\frac{7}{8}$ cup
15. $203,250
16. $218,188
17. $22,402
18. I and III

Sample Question 1

Calculate the average by adding the grades together and dividing by 9, the number of tests:

$$\text{Average} = \frac{95 + 72 + 94 + 72 + 91 + 88 + 94 + 20 + 93}{9} = \frac{719}{9} \approx 80$$

Average = 80

Calculate the median by ordering Dani's 9 grades from smallest to greatest and then identify the center number:

Median = 20, 72, 72, 88, **91,** 93, 94, 94, 95
Median = 91

Calculate the mode by finding the number that occurred most frequently:

(20, **72**, **72**, 88, 91, 93, 94, 94, 95)
Mode = 72

Dani's average is 80, median is 91, and mode is 72. The median seems to be most representative of Dani's test scores because only 3 out of the 9 scores are more than 4 points away from 91. The average is not the most representative since the test score of 20 is an outlier that brings it down significantly. (Without the 20, Dani's average goes up to an 87.)

Sample Question 2

Average (Mean)

Use the number of students achieving each score as a weight:

$$\textit{Average} = \frac{(1 \times 100) + (3 \times 95) + (6 \times 90) + (8 \times 85) + (5 \times 80) + (4 \times 75) + (2 \times 70) + (1 \times 0)}{30}$$

$$= \frac{4{,}445}{30} = 81.5$$

Even though one of the scores is 0, it must still be accounted for in the calculation of the average.

Median

Since the table is already arranged from high to low, we can determine the median merely by locating the middle score. Since there are 30 scores represented in the table, the median is the average of the 15th and 16th scores, which are both 85. Thus, the median is **85**. Even if the bottom score, 0, were significantly higher, say 80, the median would still be 85. However, the mean would be increased to 84.1. The single peculiar score, 0, makes the median a better measure of central tendency than the mean.

Mode

Just by scanning the table, we can see that more students scored an 85 than any other score. Thus, **85** is the mode. It is purely coincidental that the median and mode are the same.

CHAPTER

14 Probability

The study of mathematics, like the Nile, begins in minuteness but ends in magnificence.

—Charles Caleb Colton, English writer (1780–1832)

CHAPTER SUMMARY

This chapter introduces probability, which is something used in the business world every day. You will learn how to determine the probability of a single event happening, learn how to determine the probability of two or more independent events happening, and learn about how to calculate the probability of something *not* happening.

You've probably heard statements like, "The chances that I'll win that are one in a million," spoken by people who doubt their luck. The phrase "one in a million" is a way of stating the probability, or the likelihood, that an event will occur. Although many of us have used exaggerated estimates like this before, this chapter teaches you how to calculate probability accurately. Finding answers to questions like "What is the probability that I will draw an ace in a game of poker?" or "How likely is it that my name will be drawn as the winner of that vacation for two?" will help you decide whether the probability of an event happening is favorable enough for you to take a chance.

Determining the Probability of an Event

Probability is written as a fraction, or ratio, that compares the *number of favorable outcomes possible* to the *total number of outcomes possible.* The notation used to express "the probability of an event happening" is P(event):

$$P(Event) = \frac{\textit{Number of favorable outcomes}}{\textit{Total number of possible outcomes}}$$

Example

When you toss a coin, there are two possible outcomes: *heads* or *tails.* The probability of tossing heads is therefore 1 out of the 2 possible outcomes:

$$P(Heads) = \frac{1}{2} \begin{array}{l} \leftarrow \text{Number of favorable outcomes} \\ \leftarrow \text{Total number of possible outcomes} \end{array}$$

Similarly, the probability of tossing tails is also 1 out of the 2 possible outcomes. This probability can be expressed as a fraction, $\frac{1}{2}$, or as a decimal, 0.5. You can also think of the probability of an event as the "percent chance" that the event will occur. So, if the probability is 0.5, we can also say that it has a 50% chance of occurring. Since tossing heads and tails has the same probability, the two events are *equally likely.*

The probability that an event will occur is always a value between 0 and 1:

- If an event is a sure thing, then its probability is 1.
- If an event *cannot occur under any circumstances*, then its probability is 0.

For example, the probability of picking a black marble from a bag containing only black marbles is 1, while the probability of picking a white marble from that same bag is 0.

An event that is rather unlikely to occur has a probability close to zero. Conversely, an event that's quite likely to occur has a probability close to 1.

Example

During the holiday season, a clothing boutique decides to open seven days a week at 9 A.M. instead of 1 P.M. Every sales associate will need to work a single 9 A.M. to 1 P.M. morning shift per week during the holiday season. Karl is taking weekend acting classes and he hopes that his boss doesn't ask him to work a weekend shift. Calculate the probability that Karl will have to work a weekend shift.

$$P(Weekend) = \frac{2 \leftarrow \text{Number of weekend days}}{7 \leftarrow \text{Total number of possible outcomes}}$$

$$P(Weekday) = \frac{5 \leftarrow \text{Number of weekdays}}{7 \leftarrow \text{Total number of possible outcomes}}$$

The probability of Karl being selected to work a weekend day is $\frac{2}{7}$ since there are 2 weekend days in every week and 7 days in total. $\frac{2}{7} = 0.286$, so we can say there is a 28.6% chance that Karl will have to work on Saturday or Sunday. Similarly, the probability of Karl being selected to work a weekday is $\frac{5}{7}$ since there are 5 weekdays in every week and 7 days in total. $\frac{5}{7} = 0.714$, so we can say there is a 71.4% chance that Karl will have his extra shift on Monday through Friday. The probability of having to work a weekday

(0.714) is closer to 1 and therefore more likely than the probability of having to work a weekend day (0.286), which is closer to 0. Hopefully Karl won't have to miss his acting class!

Sometimes you might see probability written as a percentage: *There is a 60% chance of thunderstorms tomorrow.* What this means is $\frac{60}{100}$ or $\frac{3}{5}$.

Practice

Jeannie's Kettle Corn contains prizes in each bag of delicious Kettle Corn. Jeannie hopes that these prizes will encourage her customers to buy more popcorn in order to help her business grow. The following information is printed on each bag:

Collect All 4 Fabulous Prizes!

For Every 50 Boxes:
5 boxes have a Free Bag coupon for Jeannie's Kettle Corn
10 boxes have a 50% Off coupon for Jeannie's Kettle Corn
15 boxes contain one of Jeannie's Caramel Candies
20 boxes have a Glow-in-the-Dark Popcorn sticker

_____ **1.** A consumer has the highest probability of getting which prize?

_____ **2.** A consumer has the lowest probability of getting which prize?

3. Find the probability that the following prizes will be in a bag of Jeannie's Kettle Corn:

_____ **a.** a Free Bag coupon for Jeannie's Kettle Corn

_____ **b.** a 50% Off coupon for Jeannie's Kettle Corn

_____ **c.** one of Jeannie's Caramel Candies

_____ **d.** a Glow-in-the-Dark Popcorn sticker

_____ **e.** either of the two possible coupons

_____ **f.** a 50% coupon, Caramel Candy, or a Popcorn sticker

_____ **g.** anything other than the a Glow-in-the-Dark Popcorn sticker

Probability of Combined Outcomes

Sometimes it is necessary to find the probability that a combination of different events will happen. For example, a wedding planner in Seattle might want to try to calculate the probability that the first weekend in May will have temperatures higher than 60 degrees *and* will not have steady rains, before recommending that her client plan an outdoor wedding for that weekend. Weather is a little complicated to predict, though, so let's begin with a common scenario used in modeling the probability of combined outcomes: the rolling of two dice.

Example

A pair of dice is tossed at the same time and the score a player receives is the sum of both dice. What is the probability of a player receiving a score of 9?

Solution

1. Make a table showing all the possible outcomes (sums) of tossing the dice:

		Die #1					
		1	**2**	**3**	**4**	**5**	**6**
Die #2	**1**	2	3	4	5	6	7
	2	3	4	5	6	7	8
	3	4	5	6	7	8	9
	4	5	6	7	8	9	10
	5	6	7	8	9	10	11
	6	7	8	9	10	11	12

2. Determine the number of favorable outcomes by counting the number of times the sum of 9 appears in the table: **4 times.**
3. Determine the total number of possible outcomes by counting the number of entries in the table: **36.**
4. Substitute 4 favorable outcomes and 36 total possible outcomes into the probability formula, and then reduce:

$$P(Event) = \frac{\textit{\# of favorable outcomes}}{\textit{total \# of possible outcomes}}$$

$$P(\text{sum of dice is } 9) = \frac{4}{36} = \frac{1}{9}$$

Getting a sum of a 9 doesn't appear to be extremely likely with its probability of $\frac{1}{9}$.

Example

Using the rules of the previous game, what is the probability of a player receiving a score of less than 5?

Solution

1. Using the preceding table, add up all the dice tossing combinations that have a sum of 2, 3, or 4: **6 times.**
2. Since it has already been determined that there are 36 possible outcomes, substitute 6 favorable outcomes into the probability formula:

$$P(Event) = \frac{\text{\# favorable outcomes}}{\text{total \# of possible outcomes}}$$

$$P(Score\ less\ than\ 5) = \frac{6}{36} = \frac{1}{6}$$

So there is a $\frac{1}{6}$ chance that a player will score less than 5.

Try these sample questions based on throwing a pair of dice. Use the previous table to assist you. Step-by-step solutions are provided at the end of the lesson.

Sample Questions 1 and 2

What is the probability of getting a sum of *at least 7*?

What is the probability of getting a sum of 7 or 11?

Practice

A deck of ten cards contains one card with each number:

[1] [2] [3] [4] [5] [6] [7] [8] [9] [10]

_____ **4.** One card is selected from the deck. Find the probability of selecting each of the following:

_____ **a.** an odd number

_____ **b.** an even number

_____ **c.** a number less than 5

_____ **d.** a number greater than 5

_____ **e.** a 5

_____ **f.** a number less than 5, greater than 5, or equal to 5

_____ **g.** a number less than 10

_____ **h.** a multiple of 3

_____ **5.** A card will be randomly drawn and then returned to the deck. Then another card will be randomly drawn (possibly the first card again), and the two resulting numbers will be added together. Find the probability that their sum will be one of the following.
[Hint: Make a table showing the first and second cards selected, similar to the dice table used for the sample questions.]

		Card #1					
		1	**2**	**3**	**4**	**5**	**6**
Card #2	**1**	2	3	4	5	6	7
	2	3	4	5	6	7	8
	3	4	5	6	7	8	9
	4	5	6	7	8	9	10
	5	6	7	8	9	10	11
	6	7	8	9	10	11	12

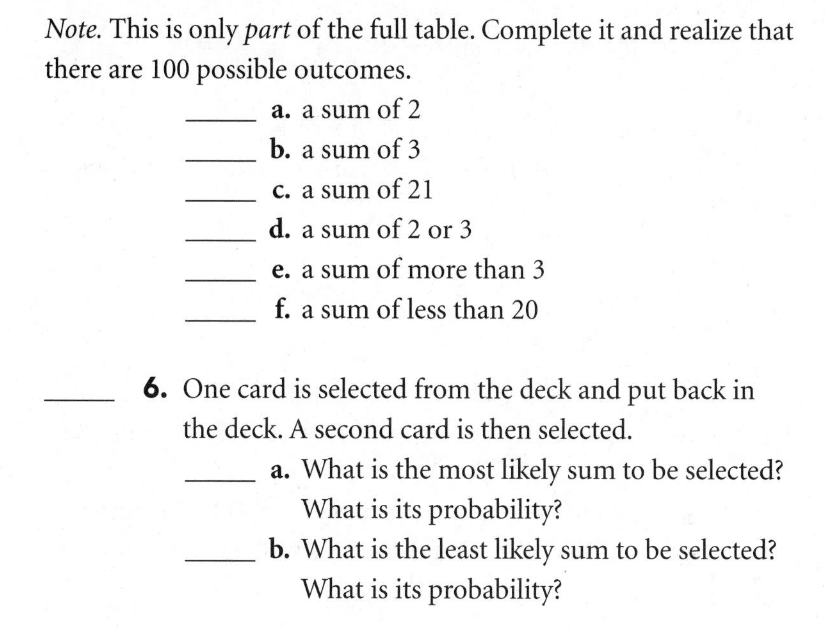

Note. This is only *part* of the full table. Complete it and realize that there are 100 possible outcomes.

_____ **a.** a sum of 2

_____ **b.** a sum of 3

_____ **c.** a sum of 21

_____ **d.** a sum of 2 or 3

_____ **e.** a sum of more than 3

_____ **f.** a sum of less than 20

_____ **6.** One card is selected from the deck and put back in the deck. A second card is then selected.

_____ **a.** What is the most likely sum to be selected? What is its probability?

_____ **b.** What is the least likely sum to be selected? What is its probability?

Probabilities That Always Sum to 1

Remember the example where Karl was going to be chosen to work a morning shift on either a weekday or a weekend? The probability that he was going to have to work a weekday was $\frac{5}{7}$ and the probability that he was going to have to work a weekend shift was $\frac{2}{7}$. Notice that the sum of these two probabilities is 1: $\frac{5}{7} + \frac{2}{7} = \frac{7}{7} = 1$. This demonstrates a very important fact about probabilities:

The sum of all the probabilities of every possible outcome of a situation is always 1.

Notice that being selected to work a weekend shift is equivalent to the probability of NOT being selected to work a weekday shift:

$$P(Weekend\ shift) = \frac{2}{7} \begin{matrix}\leftarrow \\ \leftarrow\end{matrix} \frac{\#\text{ of weekend days}}{\text{total }\#\text{ of days}}$$

$$P(NOT\ Weekday\ shift) = \frac{2}{7} \begin{matrix} \leftarrow \\ \leftarrow \end{matrix} \frac{\#\text{ of NOT weekdays}}{\text{total }\#\text{ of days}}$$

Furthermore, notice that the probability of NOT getting a weekday shift plus the probability of getting a weekday shift is equal to 1:

$$P(NOT\ Weekday\ shift) + P(Weekday\ shift) = 1$$

$$\frac{2}{7} \quad + \quad \frac{5}{7} \quad = 1$$

This demonstrates another very important fact about probabilities:

The sum of the probability that an event WILL occur and the probability that the same event will NOT occur is always 1.

This principal about probabilities is represented in the following equation:

$$\boldsymbol{P(Event\ WILL\ occur) + P(Event\ WILL\ NOT\ occur) = 1}$$

When there are only two outcomes possible, *A* or *B*, the probability that outcome *A* will occur is the same as 1 – *P*(Event *B*).

Example

A bag contains green chips, purple chips, and yellow chips. The probability of picking a green chip is $\frac{1}{4}$ and the probability of picking a purple chip is $\frac{1}{3}$. What is the probability of picking a yellow chip? If there are 36 chips in the bag, how many are yellow?

Solution

1. The sum of all the probabilities is 1: $P(\text{green}) + P(\text{purple}) + P(\text{yellow}) = 1$
2. Substitute the known probabilities: $\frac{1}{4} + \frac{1}{3} + P(\text{yellow}) = 1$
3. Solve for P(yellow): $\frac{7}{12} + P(\text{yellow}) = 1$
   The probability of picking a yellow chip is $\frac{5}{12}$. $\frac{7}{12} + \frac{5}{12} = 1$
4. Thus, $\frac{5}{12}$ of the 36 chips are yellow: $\frac{5}{12} \times 36 = \mathbf{15}$

Thus, there are 15 yellow chips.

Practice

These word problems illustrate some practical examples of probability in everyday life.

_____ **7.** A bag of jelly beans has 10 yellow, 19 purple, 24 blue, 17 pink, 29 red, and 21 green jelly beans. What's the probability that when Ben picks a jelly bean out of the bag, it will be either purple or green?

_____ **8.** A start-up company has 24 employees. 3 have their PhD, 12 have a Masters degree, 7 have an undergraduate degree, and 2 employees just graduated high school. If one employee is selected at random to give the welcome speech at a board meeting, what is the probability that the speaker will have a PhD or Masters degree?

_____ **9.** Lindsay and Jordan have teamed up to do an oral Spanish test together. Lindsay knows there is a $\frac{5}{8}$ chance that she'll draw a word that she knows and Jordan knows that there is a $\frac{3}{8}$ chance that he'll draw a word that he knows. Does this guarantee that if there are 8 words on the Spanish test, there will be a 100% chance that either Lindsay or Jordan will know a word selected at random? Justify your answer with an example or illustration.

_____ **10.** If two dice are rolled, what is the probability that the outcome will be doubles?

_____ **a.** $\frac{7}{36}$

_____ **b.** $\frac{1}{6}$

_____ **c.** $\frac{1}{3}$

_____ **d.** $\frac{1}{4}$

_____ **e.** $\frac{1}{8}$

_____ **11.** A current study shows that the probability that a woman from the United States will develop breast cancer during her life is $\frac{1}{8}$. If a group of 320 high-school-aged women are part of a lifetime study to investigate how lifestyle habits influence breast cancer, approximately how many women in this group are likely to develop breast cancer at some point?

_____ **12.** Western Pacific Clearinghouse is having a drawing at its New Year's Eve party. It is posted that the probability of winning the grand prize trip to Belize is $\frac{1}{100}$, the probability of winning a first place prize of a $1,000 flight voucher is $\frac{1}{50}$, the probability of winning a second place prize of a $50 Best Buy gift card is $\frac{1}{20}$, and the probability of winning a consolation prize of a $10 iTunes card is $\frac{1}{10}$. What is the probability that if Tina buys a ticket for this drawing, she will not win *any* prize?

Probability of Two Events Happening Together

Based on what you already have learned, it is probably easy for you to agree that the probability of rolling a 6 on a standard die is $\frac{1}{6}$ But what is the probability of rolling a 6 two times in a row? (Make a guess and write it down.) Or what about the probability of rolling a 6 with a die *three* times in a row? (Make another guess and write it down, too.) This concept of wanting two or more events to happen concurrently is called *compound probability.*

In this section we are going to look only at the compound probability of two *independent events* happening. Two events are said to be *independent*, if the outcome of one event does not have any influence over the probability of the second event happening. For example, if you roll a 6, and then pick up the die to roll it again, your probability of the second roll being a 6, is *still* $\frac{1}{6}$. The probability of two independent events happening is the *product* of both of the independent probabilities. This is written as:

$$P(\textit{Event \#1 and Event \#2}) = P(\textit{Independent Event \#1}) \times P(\textit{Independent Event \#2})$$

Let's apply this rule to our example of rolling two 6's on a die in a row:

$$P(\textit{Rolling two 6's}) = P(\textit{Rolling a 6}) \times P(\textit{Rolling a 6})$$

$$P(\textit{Rolling two 6's}) = \frac{1}{6} \times \frac{1}{6} = \frac{1}{36}$$

So the chances of rolling two 6's in a row is 1 out of 36. How close was the guess you made? We can also calculate the probability of rolling *three* 6's in row by:

$$P(\textit{Rolling three 6's}) = P(\textit{Rolling a 6}) \times P(\textit{Rolling a 6}) \times P(\textit{Rolling a 6})$$

$$P(\textit{Rolling three 6's}) = \frac{1}{6} \times \frac{1}{6} \times \frac{1}{6} = \frac{1}{216}$$

Therefore the likelihood of rolling *three* 6's in a row is 1 out of 216. That means if you did 216 trials of 3 rolls of a die, it is likely that you would only roll three 6's in *one* of those trials!

Sample Question 3

Jamie is going to roll a die two times and then flip a coin. What is the probability that she will roll a 3, then roll a number greater than 4, and then flip a tail?

Practice

For questions 13 through 17, use the following:

In a small European country, license plates are generated at random and consist of six characters. The first three characters are capital letters from the 26-letter alphabet. The last three characters are numbers, ranging from 0 to 9. A sample plate could read "ABC 123."

_____ **13.** What is the probability that the first three letters will be all vowels? (Do not count "y" as a vowel for this problem.)

_____ **14.** What is the probability that the first three letters will read "CAR"?

_____ **15.** What is the probability that all three numbers will be odd?

_____ **16.** What is the probability that the three numbers will be "777"?

_____ **17.** What is the probability that the license plate will read "CAR 777"?

TRY THIS

Gather 6 pennies and put a small piece of tape on each penny and color the tape so that you have 3 black, 2 red, and 1 blue. Put them in a bowl. You are going to run an experiment that will consist of 36 quick trials. For each trial you will pull out a penny, note its color, replace it in the bowl, and pull out a penny again. Before starting your experiment, use the probability formulas from this lesson to calculate the probabilities of pulling out 2 black pennies in a row, 2 red pennies in a row, 2 blue pennies in a row, and 2 green pennies in a row and record them in the table (keep all probabilities out of 36—do not reduce them).

COLOR	CALCULATED PROBABILITY	EXPERIMENT COUNT
2 Black Pennies in a row		
2 Red Pennies in a row		
2 Blue Pennies in a row		
2 Different Colored Pennies	(Hint: 1 – all the other probabilities)	

Now, do 36 trials and record how many times each of the different 2-penny combinations occurred in the last column of the table. (Remember to replace the first penny into the bowl before you pick the second penny!) Next, write your experimental counts over the total number of trials performed (36). This is now called your experimental probability. See how closely your experimental probabilities compare with the calculated probabilities. If you do 36 more trials, and write your results over 72 trials, you should notice that the experimental probabilities will be closer to your calculated probabilities. (If you are interested in why this is the case, you can look up The Law of Large Numbers.)

Answers

Practice Problems

1. Glow-in-the-Dark Popcorn sticker

2. Free Bag coupon for Jeannie's Kettle Corn

3. **a.** $\frac{5}{50} = \frac{1}{10}$

b. $\frac{10}{50} = \frac{1}{5}$

c. $\frac{15}{50} = \frac{3}{10}$

d. $\frac{20}{50} = \frac{2}{5}$

e. $\frac{15}{50} = \frac{3}{10}$

f. $\frac{35}{50} = \frac{7}{10}$

g. $\frac{30}{50} = \frac{3}{5}$

4. **a.** $\frac{1}{2}$ or 0.5

b. $\frac{1}{2}$ or 0.5

c. $\frac{2}{5}$ or 0.4

d. $\frac{1}{2}$ or 0.5

e. $\frac{1}{10}$ or 0.1

f. $\frac{10}{10}$ or 1

g. $\frac{9}{10}$ or 0.9

h. $\frac{3}{10}$ or 0.3

5. **a.** $\frac{1}{100}$

b. $\frac{1}{50}$

c. 0

d. $\frac{3}{100}$

e. $\frac{97}{100}$

f. $\frac{99}{100}$

6. **a.** 11, with a probability of $\frac{1}{10}$, or 0.1

b. 2 and 20, each with a probability of $\frac{1}{100}$, or 0.01

7. $\frac{40}{120} = \frac{1}{3}$

8. $\frac{15}{24} = \frac{5}{8}$

9. If there are 8 vocabulary words and Lindsay's probability of getting one correct is $\frac{5}{8}$, then that means there are 3 words she does not know. Jordan's probability of getting a word correct is $\frac{3}{8}$, which means he knows 3 words, but this does not guarantee that he knows the exact 3 words that Lindsay does *not* know. Therefore there will not be a 100% chance that either Lindsay or Jordan will know a word selected at random.

10. $\frac{6}{36} = \frac{1}{6}$

11. 40 women in this group are likely to develop breast cancer.

12. $\frac{82}{100} = \frac{41}{50}$

13. $\frac{1}{125}$

14. $\frac{1}{17{,}576}$

15. $\frac{1}{8}$

16. $\frac{1}{1{,}000}$

17. $\frac{1}{17{,}576{,}000}$

Sample Question 1

1. Determine the number of favorable outcomes by counting the number of table entries containing a sum of at least 7:

Sum	# Entries
7	6
8	5
9	4
10	3
11	2
12	+1
	21

2. Determine the number of total possible outcomes by counting the number of entries in the table: 36.
3. Substitute 21 favorable outcomes and 36 total possible outcomes into the probability formula:

$$P(\text{sum is at least 7}) = \frac{21}{36} = \frac{7}{12}$$

Since the probability exceeds $\frac{1}{2}$, it's more likely to get a sum of at least 7 than it is to get a lower sum.

Sample Question 2

There are two ways to solve this problem.

Solution #1:

1. Determine the number of favorable outcomes by counting the number of entries that are either 7 or 11:

Sum	# Entries
7	6
11	+ 2
	8

2. You already know that the number of total possible outcomes is 36. Substituting 8 favorable outcomes and 36 total possible outcomes into the probability formula yields a probability of $\frac{2}{9}$ for throwing a 7 or 11:

$$P(\text{sum is either 7 or 11}) = \frac{8}{36} = \mathbf{\frac{2}{9}}$$

Solution #2:

1. Determine two separate probabilities—$P(7)$ and $P(11)$—and add them together:

$$P(7) = \frac{6}{36}$$
$$+\ P(11) = \frac{2}{36}$$
$$P(7) + P(11) = \frac{8}{36} = \frac{2}{9}$$

Since $P(7 \text{ or } 11) = P(7) + P(11)$, we draw the following conclusion about events that don't overlap in any way:

P**(Event A or Event B) = *****P*****(Event A) + *****P*****(Event B)**

Sample Question 3

1. Determine the probability of each individual event:

$$P(\textit{roll a 3}) = \frac{1}{6}$$
$$P(\textit{roll a number greater than 4}) = \frac{2}{6} = \frac{1}{3}$$
$$P(\textit{flip a tail}) = \frac{1}{2}$$

2. Multiply all the individual probabilities together:

$$P(\textit{All three events}) = \frac{1}{6} \times \frac{1}{3} \times \frac{1}{2} = \frac{1}{36}$$

So the probability of Jamie rolling a 3, rolling a number greater than 4 and flipping a tail is $\frac{1}{36}$. Good luck, Jamie!

CHAPTER 15

Perimeter and Circumference

Geometry existed before the creation.

—PLATO, classical Greek philosopher (427 B.C.E.–347 B.C.E.)

CHAPTER SUMMARY

Geometry is the body of math that encompasses 2-dimensional and 3-dimensional shapes. This field of math is widely applied in many different lines of work and is also helpful in problem solving in personal life. In this first chapter, we will learn about polygons, circles, perimeter, and circumference.

Perimeter is the distance around the outside of a 2-dimensional object. Perimeter is a widely used concept in many workplaces. A carpenter would use the perimeter of a room in order to determine how many feet of crown molding she needs to surround the room, a tailor would use the perimeter of a blanket to determine how much silk trim he needs to cover its edges, and a farmer would use the perimeter of a pasture in order to know how much fencing to purchase to contain livestock. But before we begin discussing how to calculate perimeter, let's review the shapes we'll be working with first: polygons.

Polygons

What Is a Polygon?

A polygon is a closed, planar (flat) figure formed by three or more connected line segments that don't cross each other. Familiarize yourself with the following polygons; they are the three most common polygons in real-life situations.

Triangle: A triangle is a polygon with three sides. You will learn later about the characteristics of special triangles.

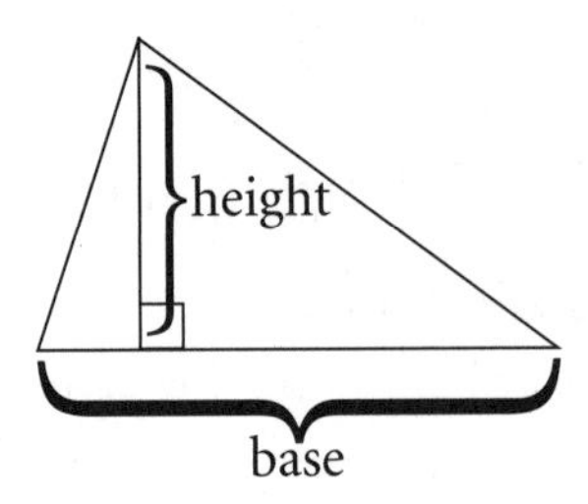

Rectangle: A rectangle is a 4-sided polygon that has four right angles. (A right angle has a measure of 90 degrees, which looks like the corner of a page in this book.) In rectangles, the opposite sides are parallel and equal in length.

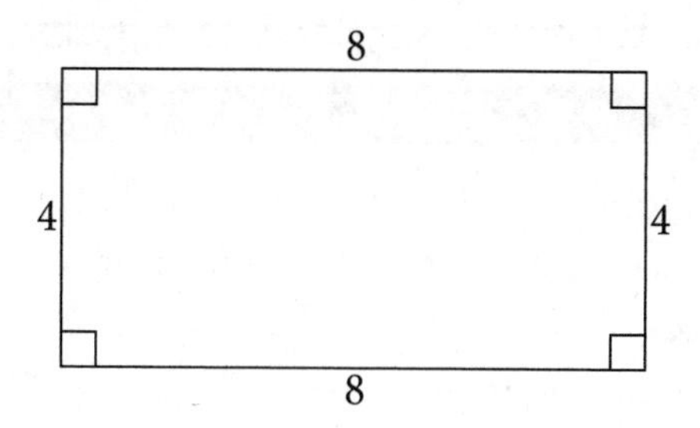

Square: A square is a 4-sided polygon that has four right angles *and* four equal sides. (A square is really a special type of rectangle that has equal sides.) In squares, the opposite sides are parallel.

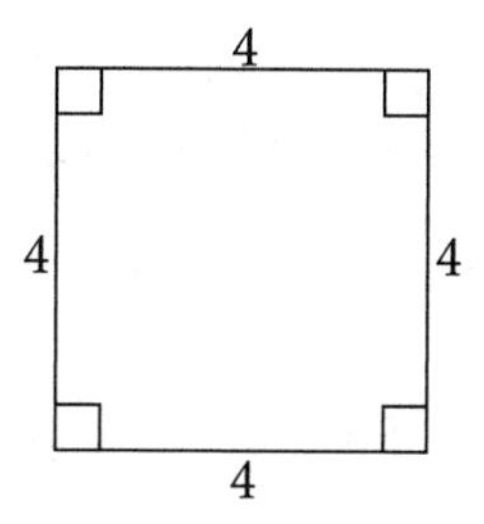

Perimeter

Perimeter is the distance around the outside edge of a polygon. The word *perimeter* is derived from *peri*, which means *around* (as in *peri*scope and *peri*pheral vision), and *meter*, which means *measure*. Thus, *perimeter* is the *measure around* a figure.

To find the perimeter of a polygon, add the lengths of the sides.

CAUTION

The most important thing to remember after finding the perimeter of a polygon, is to represent the perimeter as a unit of length, like inches, feet, yards, meters, and so on. If you just write that the perimeter of a garden plot is "20," then it's impossible for others to know whether the perimeter is 20 feet, 20 yards, 20 meters, or even 20 inches (that would be a garden plot for 1 earthworm). Therefore, make sure to always write the unit of measurement after the sum of the lengths of sides: *P = 20 yards*. If no unit of measurement is provided in an illustration, simply write the word *units* after the perimeter: *40 units*.

Example

Find the perimeter of the following polygon:

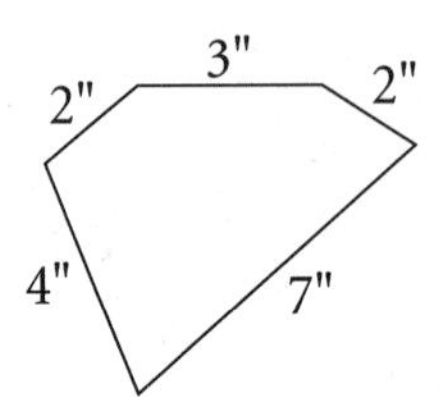

Solution

Write down the length of each side and add:

3 inches
2 inches
7 inches
4 inches
+ 2 inches
18 inches

Practice

Find the perimeter of each polygon.

1.

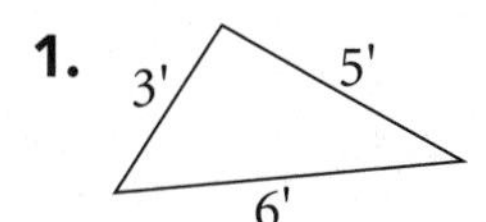

2.

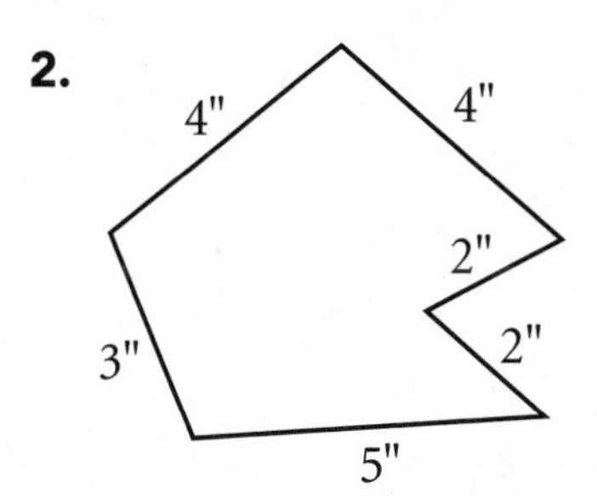

3.

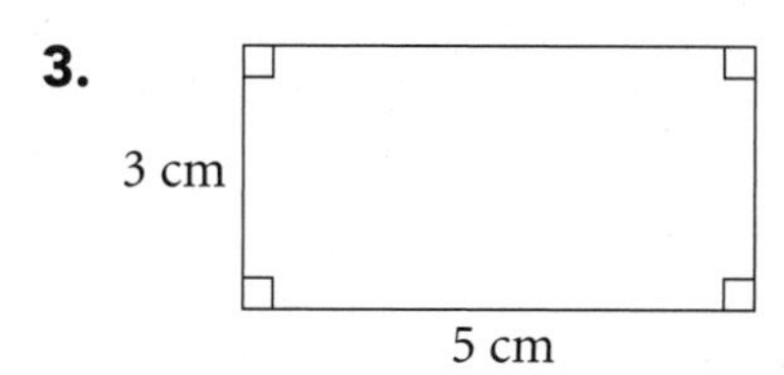

4.

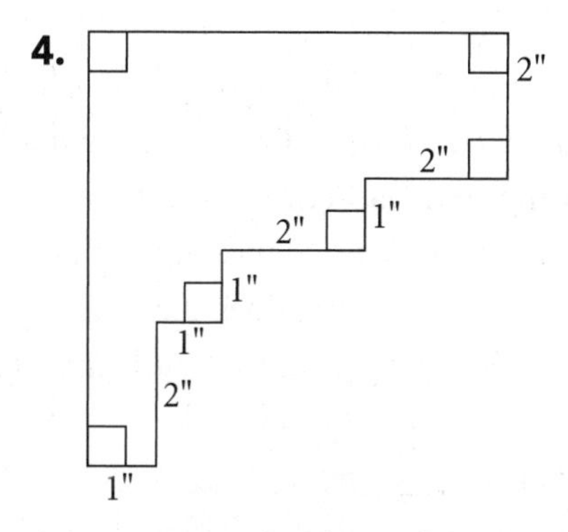

Note: The hash marks drawn in every side of the two figures in questions 5 and 6 indicate that all the sides have equal length. Look for these in future illustrations.

5.

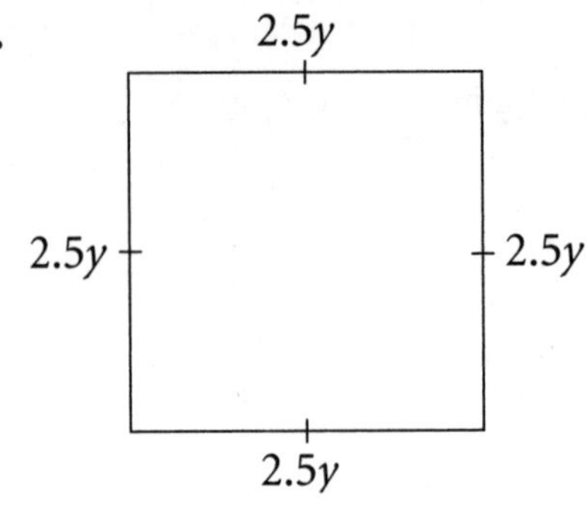

6.

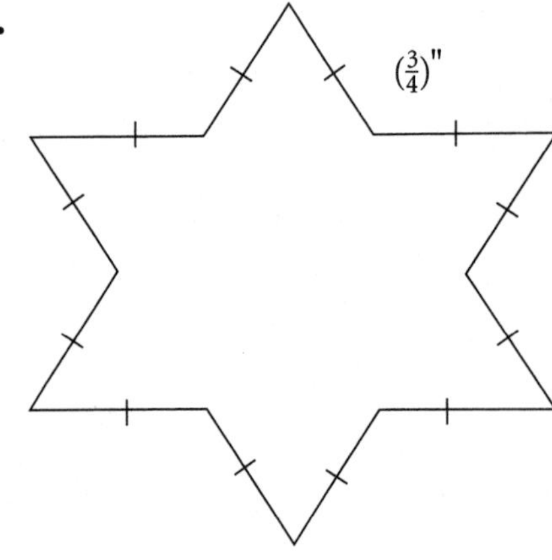

Word Problems

_____ **7.** Maryellen has cleared a 10-foot by 6-foot rectangular plot of ground for her herb garden. She must completely enclose it with a chain-link fence to keep her dog out. How many feet of fencing does she need, excluding the 3-foot wooden gate at the south end of the garden?

_____ **8.** Terri plans to hang a wallpaper border along the top of each wall in her square dressing room. Wallpaper border is sold only in 12-foot strips. If each wall is 8 feet long, how many strips should she buy?

_____ **9.** Barry is a stonemason and he has made the following geometric tile pattern for the corner of a client's yard, where the fire pit will sit. Each square tile shown is 3 feet by 3 feet. He needs to install a steel rail along the outside of all the tiles in order to keep them tightly in place. How many feet of steel rail will Barry need for this job?

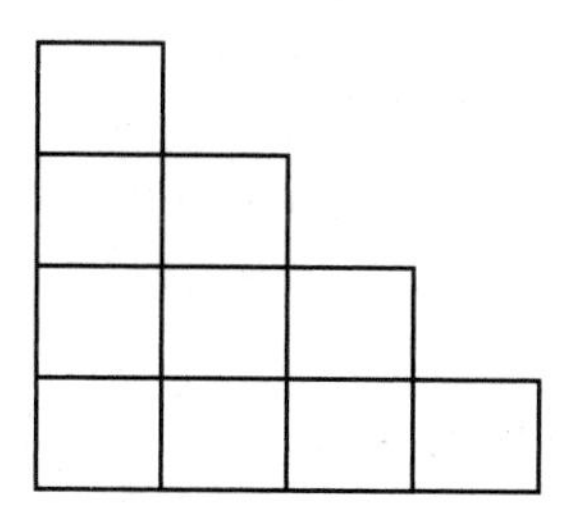

Triangles

Types of Triangles

As you already know, a triangle is a polygon with 3 *sides*. Another important fact about triangles has to do with a triangle's 3 *angles*. Regardless of the shape of a triangle, whether it's flat and narrow or tall and stately, the three angles of any triangle will always add up to 180°. This theorem is illustrated here:

The sum of the angles in a triangle is 180°:

$$\angle A + \angle B + \angle C = 180°$$

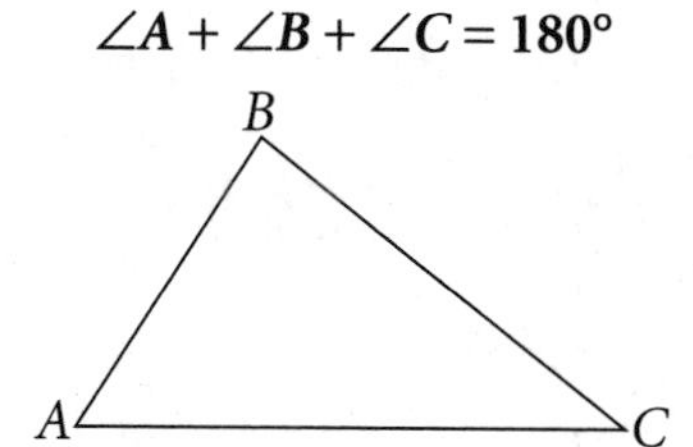

Now let's look to see how triangles are classified by either the lengths of their sides or the measures of their angles. The three types of triangles that are most important to be familiar with are:

Equilateral Triangle

- 3 equal angles, each 60°
- 3 equal sides

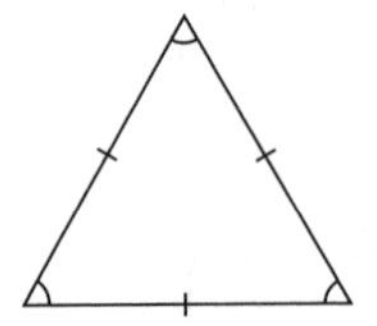

Isosceles Triangle

- 2 equal angles, called *base angles*; the third angle is the *vertex angle.*
- Sides opposite the base angles are equal.

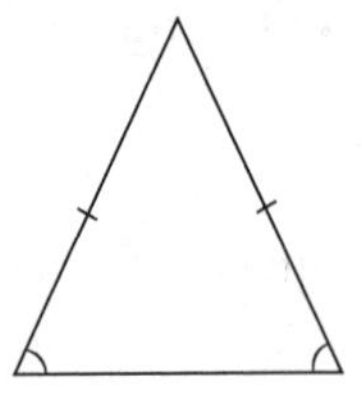

Right Triangle

- 1 right (90°) angle, the largest angle in the triangle, is marked with a little square in the corner to show that it is the right angle.
- Side opposite the right angle is the *hypotenuse*, the longest side of the triangle.
- The other two sides are called *legs*.

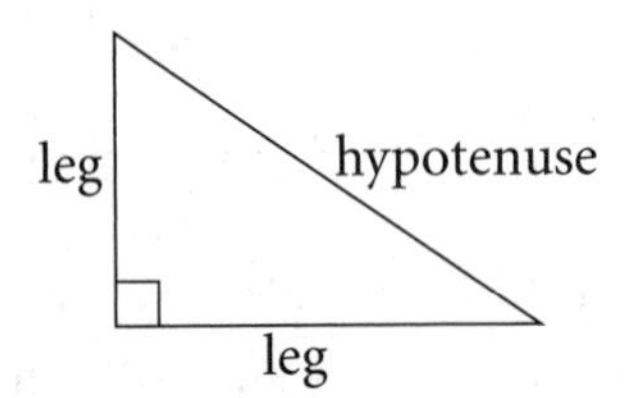

Practice

Classify each triangle as equilateral, isosceles, or right. Remember, some triangles may have more than one classification.

10.

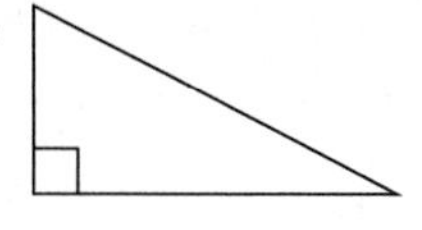

11.

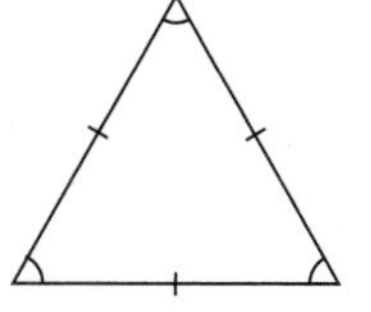

12.

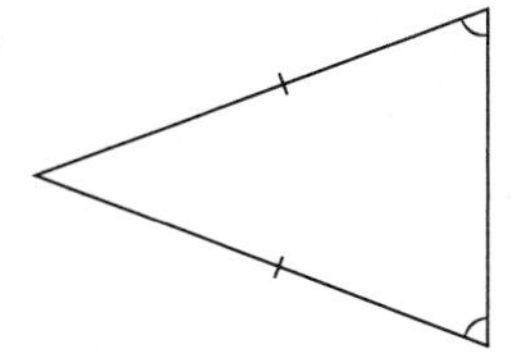

13.

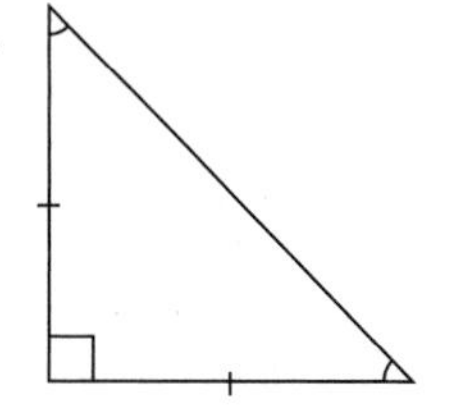

_____ **14.** A synagogue has a stained glass window that is an equilateral triangle. Frank has been hired to install wood trim around the window and he's come by the synagogue to take measurements, but he doesn't feel like carrying his heavy ladder into the synagogue. He can reach the bottom edge of the triangular window to measure it and it is 10.5 feet long. How many feet of trim will he need to order to go around the entire window?

_____ **15.** Tami used exactly 90 feet of lights to line the edge of her roofline, which has the shape of an isosceles triangle. She wants to add an additional strand of lights to the sloping edges of her roof, but she doesn't know how long each side is. If the horizontal edge of her roofline is 40 feet long, how long are each of the sloped edges?

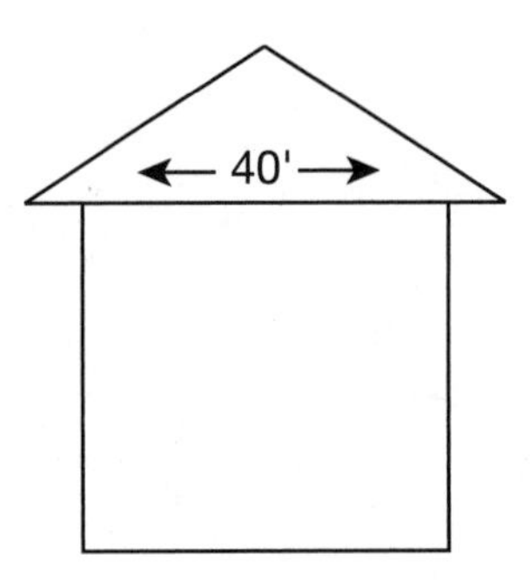

TIP

The longest side of a triangle is always opposite the largest angle, and the smallest side is opposite the smallest angle. That is why in an isosceles triangle, the two base angles are equal to and opposite of the two equal sides.

This rule implies that the second longest side is opposite the second largest angle, and the shortest side is opposite the smallest angle.

Hook

Visualize a door and its hinge. The more the hinge is open (largest angle), the fatter the person who can get through (longest side is opposite); similarly, for a door that's hardly open at all (smallest angle), only a very skinny person can get through (shortest side is opposite).

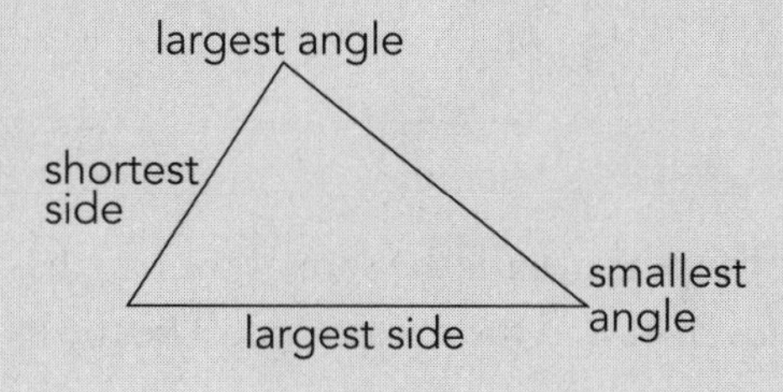

Right Triangles and the Pythagorean Theorem

Right triangles have a special theorem that can be used to determine a missing side length when the two other sides of the triangle are known. The Pythagorean theorem states that *the sum of the squares of the two legs is equal to the square of the hypotenuse* (remember that the *hypotenuse* is the longest side of a triangle, which is always opposite the right angle).

Pythagorean Theorem: $(Leg)^2 + (Leg)^2 = (Hypotenuse)^2$

Notice that in the accompanying illustration, the legs of the right triangle are labeled a and b, and the hypotenuse is labeled c. This is a common convention for how right triangles are labeled, but of course it is not required.

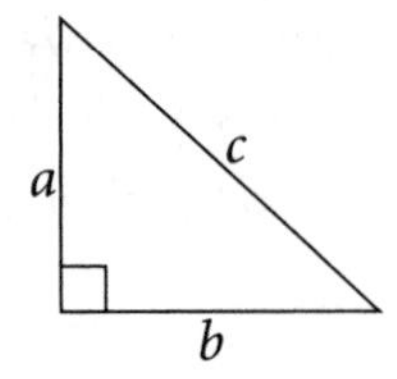

Pythagorean Theorem: $a^2 + b2 = c2$

Example

What is the perimeter of the following triangle?

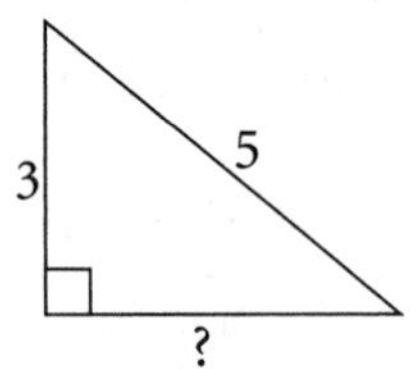

1. Since the perimeter is the sum of the lengths of the sides, we must first find the missing side. Use the Pythagorean theorem: $3^2 + b^2 = 5^2$
2. Substitute the given sides for two of the letters. Remember: Side c is always the hypotenuse: $9 + b^2 = 25$
3. To solve this equation, subtract 9 from both sides: $-9 \quad -9$

   $b^2 = 16$
4. Then, take the square root of both sides. (Note: Refer to Lesson 18 to learn about square roots.) $\sqrt{b^2} = \sqrt{16}$

Thus, the missing side has a length of 4 units. $b = \mathbf{4}$

5. Adding the three sides yields a perimeter of 12 units: $3 + 4 + 5 = \mathbf{12}$

Practice

Use the Pythagorean theorem to find the missing side length in each triangle and then calculate the perimeter of each triangle.

16.

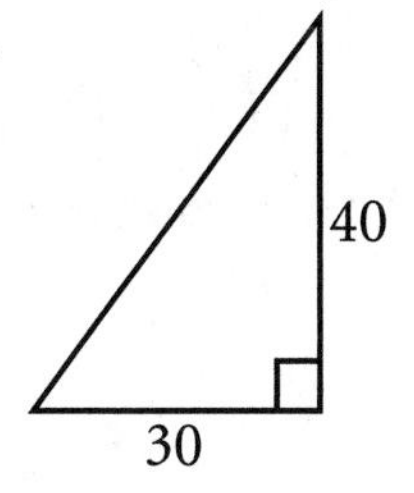

17.

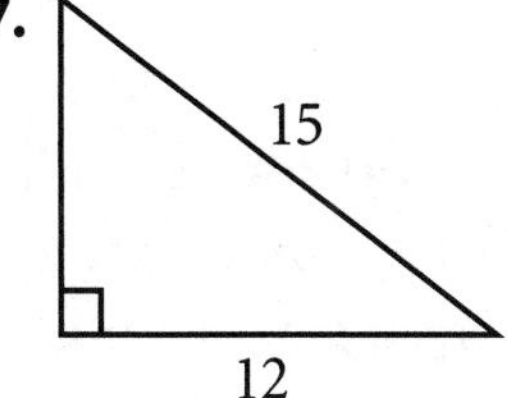

18.

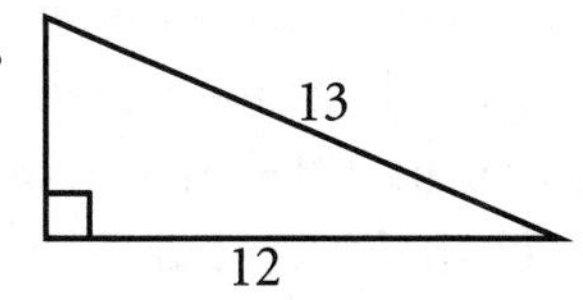

19.

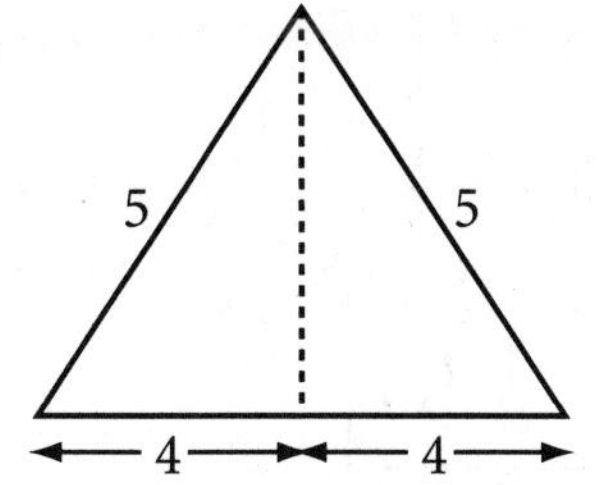

Word Problems

_____ **20.** Erin is flying a kite. She knows that 260 feet of the kite string has been let out. Also, a friend is 240 feet away and standing directly underneath the kite. How high off the ground is the kite?

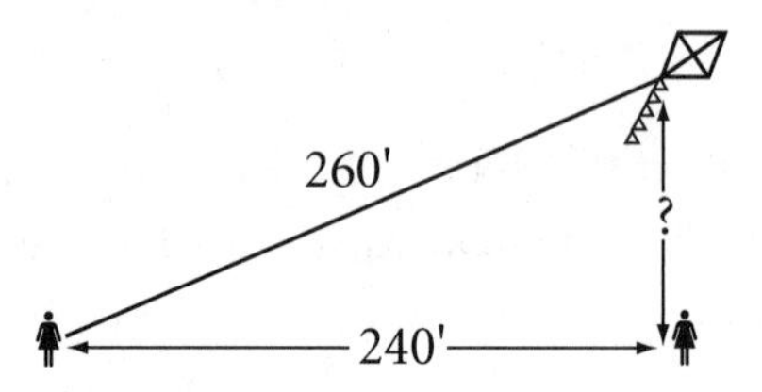

_____ **21.** What is the length of a side of an equilateral triangle that has the same perimeter as the following triangle?

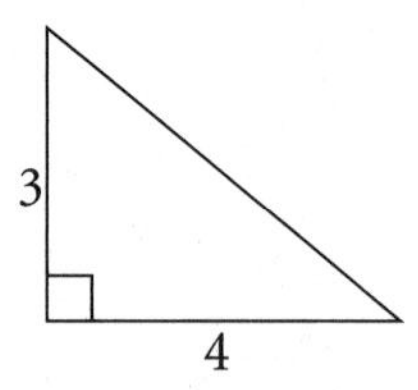

_____ **22.** Gil is a kicker for his college football team and he must successfully kick a field goal in order to win the game against their rivals. He is standing 40 yards away from the base of the field goal and the crossbar is 10 feet above the ground. What is the minimum amount of feet that Gil must kick the football in order to make it over the crossbar? (Use the following illustration and notice that two different units of measurement are given here.)

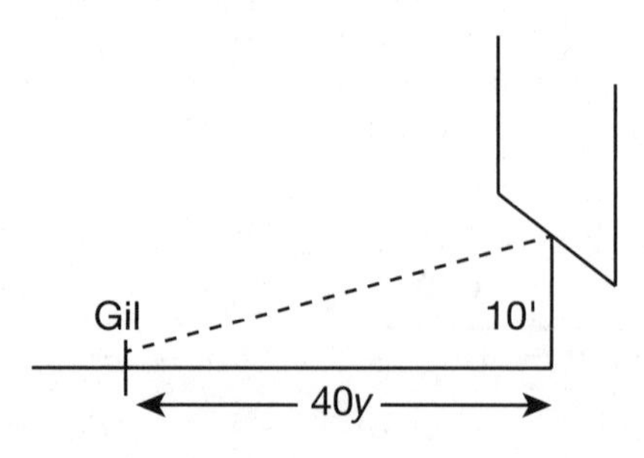

Circles

We can all recognize a circle when we see one, but its definition is more technical than that of a quadrilateral. A circle is a set of points that are all the same distance from a given point called the center.

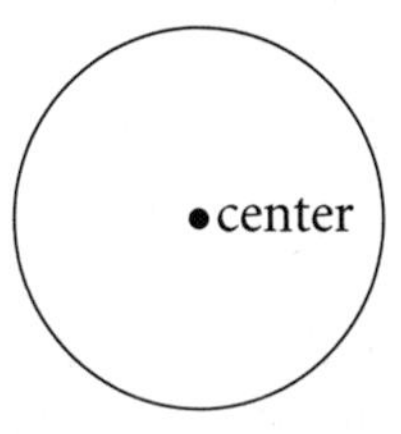

You are likely to come across the following terms when dealing with circles:

Radius: The distance from the center of the circle to any point on the circle itself. The symbol r is used for the radius.

Diameter: The length of a line that passes across a circle through the center. The diameter is twice the size of the radius. The symbol d is used for the diameter.

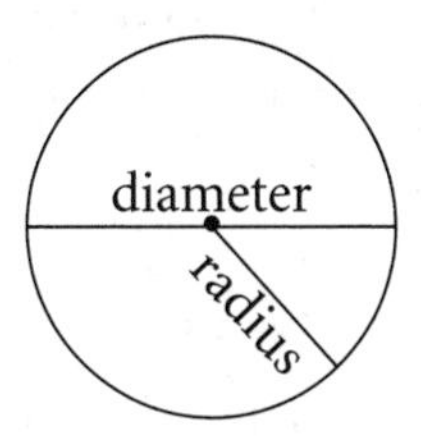

Circumference

The *circumference* of a circle is the distance around the circle (comparable to the concept of the *perimeter* of a polygon). To determine the circumference of a circle, use either of these two equivalent formulas:

$$\textit{Circumference} = 2\pi r$$

or

$$\textit{Circumference} = \pi d$$

Circumference = $2\pi r$ or Circumference = πd

The formulas should be written out as:

$$2 \times \pi \times r \text{ or } \pi \times d$$

It helps to know that:

- r is the radius
- d is the diameter
- π is approximately equal to 3.14 or $\frac{22}{7}$

Note: Letters of the Greek alphabet, like π (*pi*), are often used in math formulas. Perhaps that's what makes math seem like Greek to some people! Since a circle is shaped like a *pie*, use that fact as a way to remember that you must use π when calculating the distance around a circle.

Example

Find the circumference of a circle whose radius is 7 inches.

1. Since you are given the radius, write the radius version of the circumference formula: $C = 2\pi r$
2. Substitute 7 for the radius: $C = 2 \times \pi \times 7$
3. Depending on whether you want your final answer to be in fraction format or decimal format, substitute $\frac{22}{7}$ or 3.14 for π and multiply:
   $C = 2 \times \frac{22}{7} \times 7$; $C =$ **44**
   $C = 2 \times 3.14 \times 7$; $C =$ **43.96**
   Or, in some cases it might be appropriate to keep your answer in terms of *pi*:
   $C = 2 \times \pi \times 7$; $C =$ **14π**

All the answers—**44 inches**, **43.96 inches**, and **14π inches**—are considered correct.

Example

What is the diameter of a circle whose circumference is 62.8 centimeters? Use 3.14 for π.

1. Since you're asked to find the diameter, write the diameter version of the circumference formula:
   $C = \pi d$
2. Substitute 62.8 for the circumference, 3.14 for π, and solve the equation by dividing by $3.14 \times d$:
   $62.8 = 3.14 \times d$
   The diameter is **20 centimeters.**
   $62.8 = 3.14 \times \mathbf{20}$

Practice

23. Find the circumference:

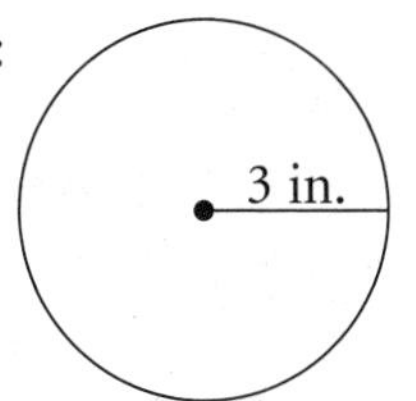

24. Find the radius and diameter:

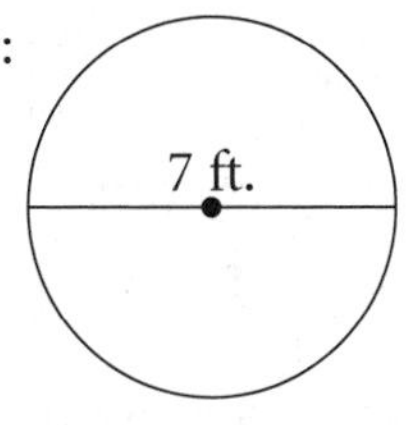

_____ **25.** If a can is 5 inches across the top, how far around is it?

_____ **26.** Find the circumference of a water pipe whose radius is 1.2 inches.

_____ **27.** What is the circumference of a circle whose radius is the same size as the side of a square with a perimeter of 20 meters?

28. Find the radius and diameter:

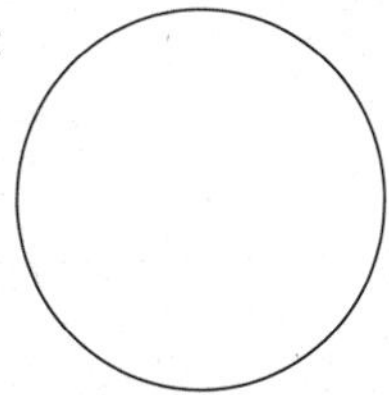

Circumference = 10π ft.

Working Backward with Perimeter and Circumference Formulas

Now that you have the formula for the circumference of a circle, let's look at the formulas that can be used for finding the perimeters of squares and rectangles. Even though it's easy to add the sides of a polygon together, formulas provide a more efficient way to solve problems quickly. These formulas can also be used to work backward to calculate the side length of a polygon or the radius or diameter of a circle when only the perimeter or circumference is given.

Squares: The four sides of a square are all equal in measure and since the side length of squares is commonly referred to as s, the perimeter of a square is $P = s + s + s + s = 4s$:

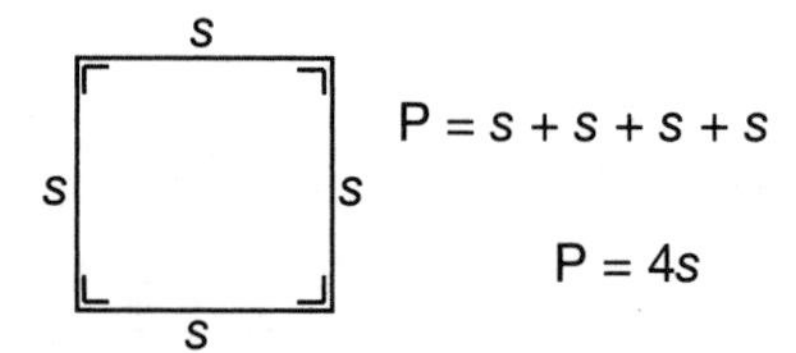

Perimeter of Square = 4*s*

Rectangle: The two longer sides of a rectangle are equal in measure and are referred to as the *lengths* of a rectangle (l). The two shorter sides of a rectangle are equal in measure and are referred to as the *widths* of a rectangle (w). Therefore, the perimeter can be written as $l + l + w + w = 2l + 2w$, which can also be written as $2(l + w)$:

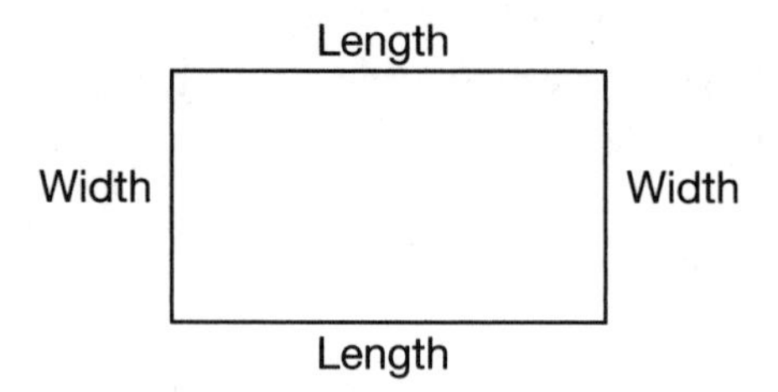

Perimeter of Rectangle = $2l + 2w$ or $2(l + w)$

Example

Santa Monica City Garden Cooperative just received a generous donation of a plot of land that is 100 feet long by 20 feet wide to use for a community garden. They have 160 feet of fencing left over from deconstructing a garden plot that was on a different plot of land that was sold to a developer. If they want to use the entire 20-foot width of the new plot, and all 160 feet of their leftover fencing to fence in a rectangular plot, how long with the fenced plot be?

Solution

1. First, write down the formula for the perimeter of a rectangle:

$$P = 2l + 2w$$

2. Next, fill in the information you were given ($P = 160$ and $w = 20$):

$$160 = 2l + 2(20)$$

3. Lastly, solve for l:

$$\begin{array}{l} 160 = 2l + 40 \\ \underline{-40 \qquad -40} \\ 120 = 2l \end{array}$$

so l must equal **60**

The new fenced garden plot will be 60 feet long by 20 feet wide.

Practice

Use the formulas and work backwards to find the missing information in each of the following problems.

Word Problems

_____ **29.** What is the length of a side of a square room whose perimeter is 58 feet?

_____ **30.** Bren is a woodworker who makes custom kitchen tables. He designs his tables so that every guest has 2 feet of linear space for "elbow room." A new customer would like a table that is exactly 10 feet long and she also wants it to be able to accommodate 16 guests. What is the perimeter needed to fit 16 guests comfortably and what is width of the table Bren must make to meet his client's needs?

_____ **31.** Giant sequoias are the world's most massive trees. If a scientist is studying a giant sequoia that measures 113 feet around, find the diameter of this particular tree. (Use 3.14 for π.)

_____ **32.** If Ryan's bicycle tire has a diameter of 24 inches, how many feet will it travel in a single rotation? How many times must it make a full rotation during his 2-mile bike commute to work? (Hint, there are 5,280 feet in a mile.)

TRY THIS

Crown molding is a wooden trim that goes along the top of a wall, where the wall meets the ceiling. When you are at home, take a look at your bedroom and guess how many linear feet of crown molding would be needed for that room. Then make a sketch of your bedroom and use a tape measure to find and label the dimensions in your drawing. Next, calculate the perimeter of your bedroom. This will represent how many feet of crown molding you would need for your room. How close was your original guess? How much would this cost you if you bought crown molding costing $3 per linear foot?

Answers

1. 14 ft.
2. 20 in.
3. 16 cm
4. 24 in.
5. 10 y
6. 9 in.
7. 29 ft. (The perimeter is 32 feet, but you must subtract 3 feet for the wooden gate.)
8. 3 strips (There will be 4 feet of border left over.)
9. 48 ft.
10. Right
11. Equilateral
12. Isosceles
13. Right and isosceles
14. 31.5 feet
15. Each edge is 25 feet so she will need 50 feet of lights.
16. $P = 120$ units
17. $P = 36$ units
18. $P = 30$ units
19. $P = 18$ units
20. 100 feet
21. 4 units. The hypotenuse of the triangle shown is 5, making its perimeter 12. Since all three sides of an equilateral triangle are the same length, the length of each side is 12 divided by 3, which is 4.
22. 120.4 feet
23. 6π in. or 18.8 in.
24. radius = 3.5 ft. and diameter = 7 ft.
25. 5π in. = 15.7 in.
26. 2.4π in. = 7.5 in.
27. 10π m = 31.4 m
28. $r = 5$, $d = 10$

29. 14.5 ft.

30. Perimeter = 32 feet, width = 6 feet

31. 36 feet

32. 6.28 feet in a single rotation; approximately 1,682 rotations are needed to travel 2 miles.

CHAPTER

16 Area

CHAPTER SUMMARY

In the previous chapter you learned how to measure the distance *around* an object and in this lesson you will learn how to measure the *area*, or space *within* a 2-dimensional polygon or circle.

Now that you're familiar with finding the perimeter and circumference of two-dimensional shapes, it's time to learn about the *area* of two-dimensional shapes. *Area* is the amount of space *within* a two-dimensional figure. The trim needed to go along the border of a room is represented by the perimeter of the room, but the number of square tiles needed to completely cover the floor of that same room is represented by area. People sometimes confuse perimeter and area, but they are very different types of measurements since perimeter is measured in linear units (meaning it could be represented as a straight line), whereas area is measured in square units. Let's take a closer look at square units.

Measuring Area in Square Units

As stated, area is the amount of space *within* a two-dimensional figure and it is calculated in *square units* of measurement. *Square units* have two equal dimensions: length and width. For instance, if you have 1 square yard of canvas, you have a piece of fabric that is 1 yard long by 1 yard wide. This can be written as "1 square yard" or "1 $yard^2$." If you are measuring something in feet, like the dimensions of a room, then the area is measured in *square feet.* One square foot is 1 foot long by 1 foot wide. Other square units you will commonly see include (but are not limited to) square inches ($in.^2$), square meters (m^2), square kilometers (km^2), and so on. When the exact unit of measurement is unknown, the area is defined in terms of square units—the following illustration is of one square unit:

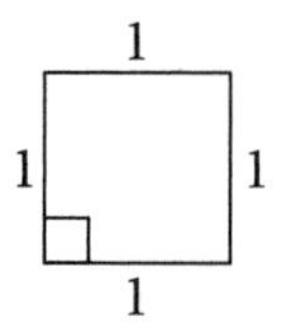

You could measure the area of any figure by counting the number of square units the figure occupies. The following two figures are easy to measure because the square units fit into them evenly.

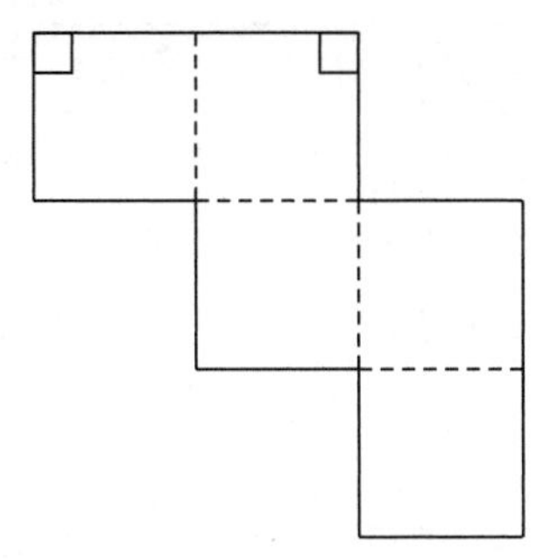

5 square units of area

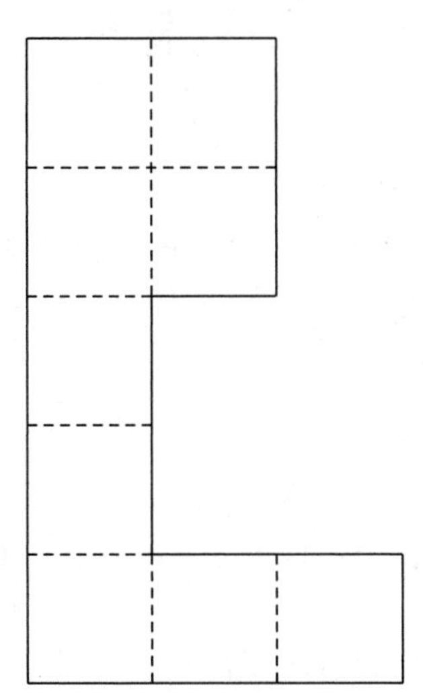

9 square units of area

Now look at the next two figures that are not as easy to measure with square units because the square units do not neatly and evenly fit into them.

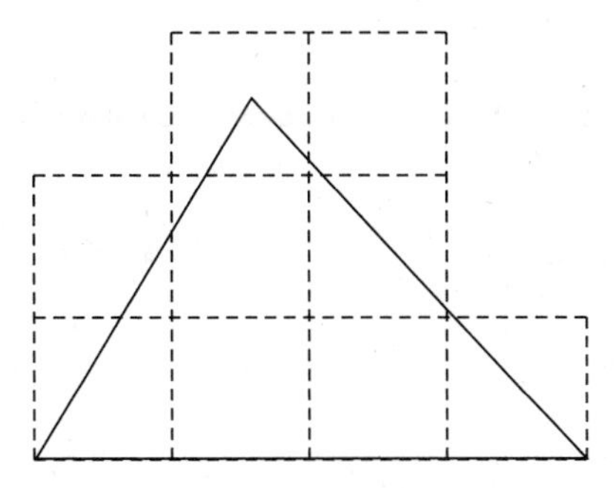

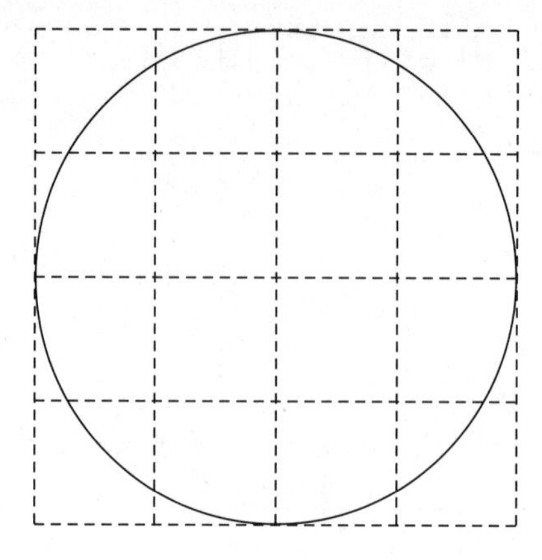

Because it's not always practical to measure a particular figure's area by counting the number of square units it occupies, an area formula is often used. As each figure is discussed, you'll learn its area formula.

Area of Rectangles and Squares

Let's say your boss has asked you the purchase new ceramic tiles for the entryway to your office building. Each tile is 1 square foot and the entryway is 5 feet long and 3 feet wide. Looking at the following illustration, can you see how many tiles you will need to order for this renovation?

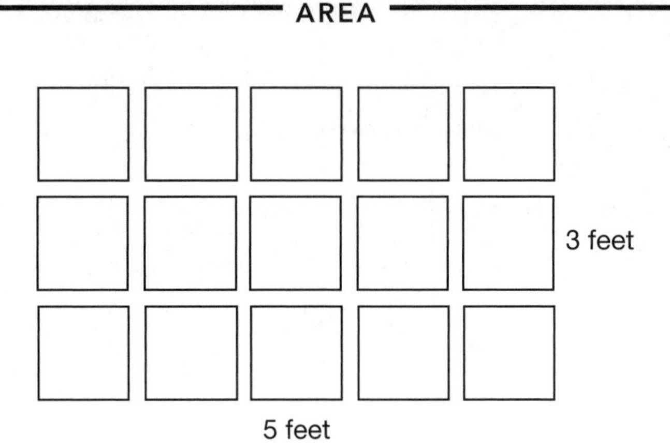

By counting all the individual tiles in the figure, you can probably see that 15 tiles will be needed for the entryway. You might have also noticed that if you multiply the length of the entryway (5 feet), by the width of the entryway (3 feet) that this also gives you the number of square tiles needed for the entryway. This technique actually uses the area formula for rectangles, which is to multiply the *length* times the *width*:

area of rectangle* = *length* × *width

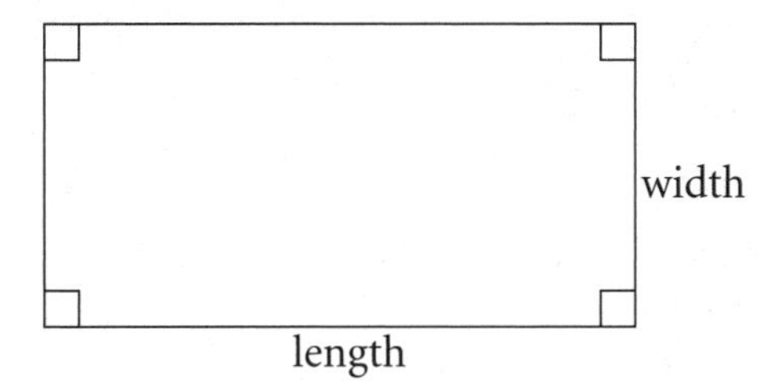

Example

Find the area of a rectangle with a base of 4 meters and a height of 3 meters.

1. Draw the rectangle as close to scale as possible.
2. Label the size of the base and height.

3. Write the area formula $A = b \times h$; then substitute the base and height numbers into it:
   $A = 4 \times 3 = \mathbf{12}$
   Thus, the area is **12 square meters.**

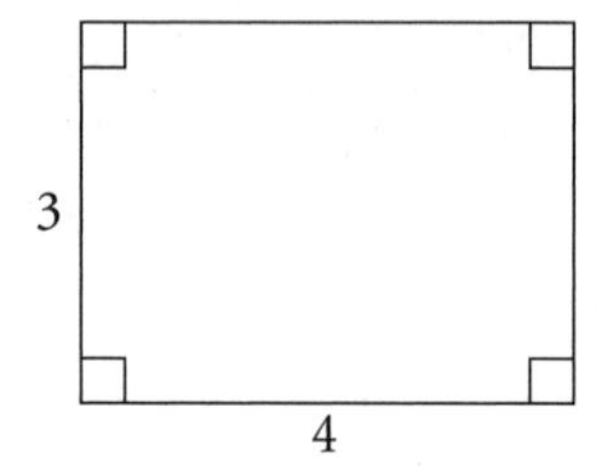

Although you can use the rectangle area formula for squares, squares have their own area formula, which is helpful if you need to work backward to determine the side length of a square with a known area. As you recall, squares have equal sides lengths that are commonly referred to as *s*:

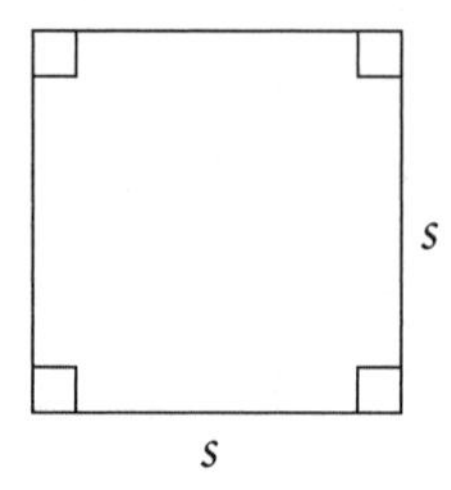

To find the area of a square, multiply the side length by itself, which will give you s^2:

$$\textbf{\textit{area of square}} = \textbf{\textit{side}} \times \textbf{\textit{side}} = \boldsymbol{s^2}$$

Example

Kayvon's client would like him to paint a large purple square on the side of her building as part of an art installation. Kayvon is considering purchasing a quart of paint, which says that it can cover an area of 85 square feet. What is the side length of the largest square that could be painted with this quart of paint? Round your answer to the nearest whole foot.

1. Write the area formula for squares: $\boldsymbol{A = s^2}$
2. Fill in the area that the paint label says it will cover: $\mathbf{85} = s^2$
3. Take the square root of both sides to solve for s: $\sqrt{85} = \sqrt{s^2}$
4. Since $\sqrt{85}$ is 9.2, $s = $ **9.2 feet** and the largest square to the nearest whole foot will be 9 feet by 9 feet.

Practice

Find the area of the following polygons or the missing side length. Make sure you represent your answer in the correct units.

_____ **1.**

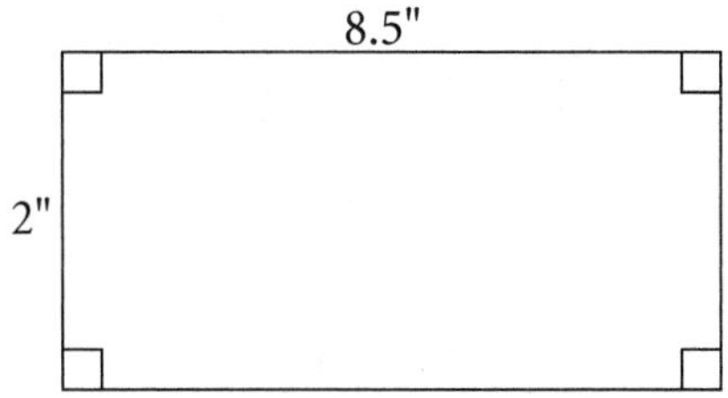

8.5"

2"

_____ **2.**

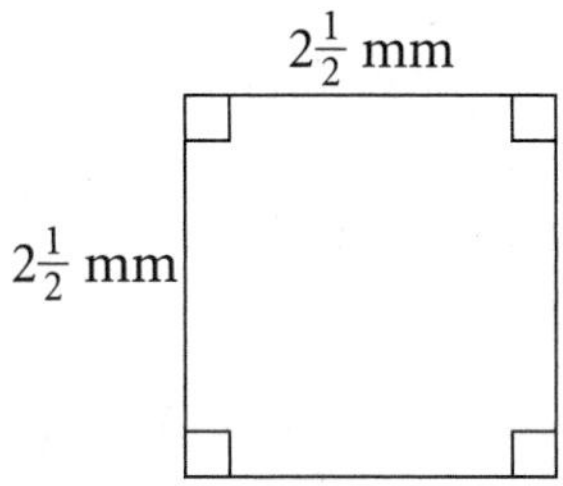

$2\frac{1}{2}$ mm

$2\frac{1}{2}$ mm

_____ **3.** Area of Rectangle = 27 $ft.^2$; find the width:

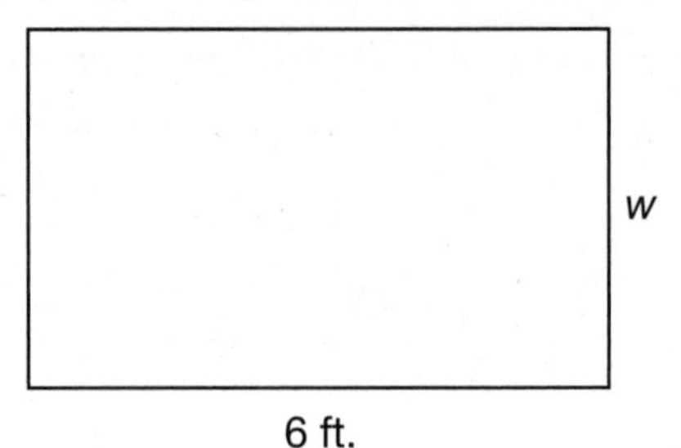

_____ **4.** Area of square = 25 y^2; find the side length:

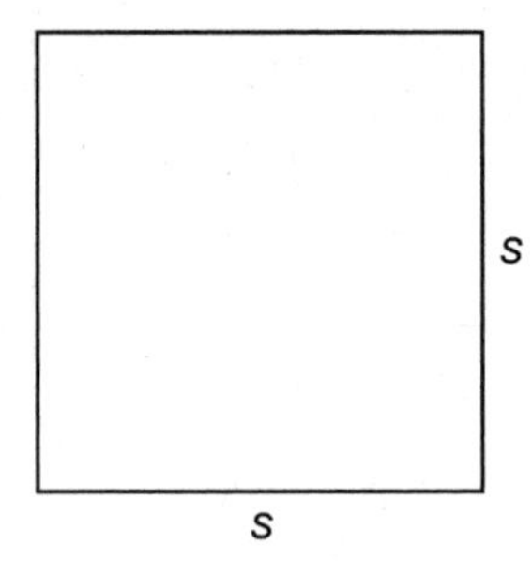

_____ **5.** One can of paint covers 200 square feet. How many cans will be needed to paint the ceiling of a room that is 32 feet long and 25 feet wide?

_____ **6.** Erin is laying square tiles that measure 24 inches by 24 inches in her outdoor patio. If her patio measures 22 feet by 18 feet, how many tiles will she need?

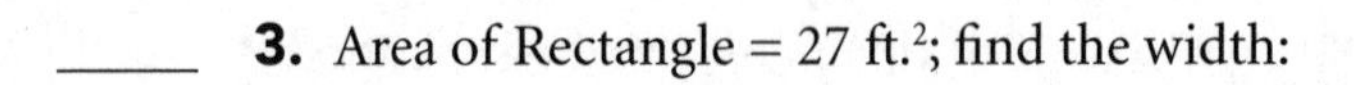

_____ **7.** Sophia is the head of maintenance at Woodier School, which is located in the high desert of Oregon. She has decided to replace all the grass on the school's property with artificial grass in order to cut down on water waste and pesticide use. The artificial grass she would like to purchase costs $3.50 per square foot. Use the dimensions in the following illustration to determine the area of the yard and the total cost of the artificial grass needed for this project.

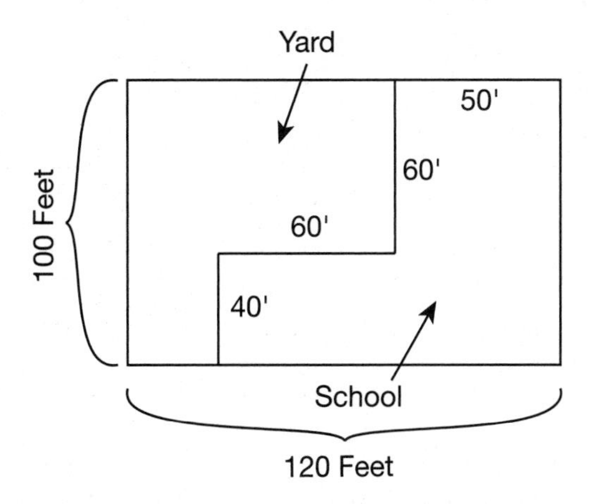

_____ **8.** Zelda has 16 feet of fencing to make a rectangular or square chicken coop. If each side of the coop will be a measurement of whole feet (no inches or partial feet), what dimensions give Zelda's chickens the most area to walk around? Fill out the following table to find your answer:

LENGTH	WIDTH	$P = 2l + 2w$	$A = l \times w$
1	7 feet	16 feet	7 ft.^{2}
2		16 feet	
3		16 feet	
4		16 feet	
5		16 feet	
6	2 feet	16 feet	12 ft.^{2}
7		16 feet	
8		16 feet	

_____ **9.** Using the table you completed in question 8, when given a fixed perimeter, describe the shape that yields the coop with largest area and describe the shape that yields the coop with the smallest area.

_____ **10.** Cinder blocks are the main construction material used to build houses in Guatemala. If you were designing a house in Guatemala City to accommodate a family of four and wanted to keep the material costs down to a minimum, what would be the most cost-efficient shape for the house? Use your findings from question 9 and justify your answer.

Area of Triangles

To find the ***area*** of a triangle, use this formula:

$$\textbf{\textit{area}} = \tfrac{1}{2}\,(\textbf{\textit{base}} \times \textbf{\textit{height}})$$

Although any side of a triangle may be called its ***base***, it's often easiest to use the side on the bottom. To use another side, rotate the page and view the triangle from another perspective.

A triangle's ***height*** (or ***altitude***) is represented by a perpendicular line drawn from the angle opposite the base to the base. Depending on the triangle, the height may be inside, outside, or on the legs of the triangle. Notice the height of the second triangle: We extended the base to draw the height perpendicular to the base. The third triangle is a ***right*** triangle: One leg may be its base and the other its height.

> **CAUTION**
>
> The height must be perpendicular to the base. As such, it might NOT be one of the actual sides of the triangle.

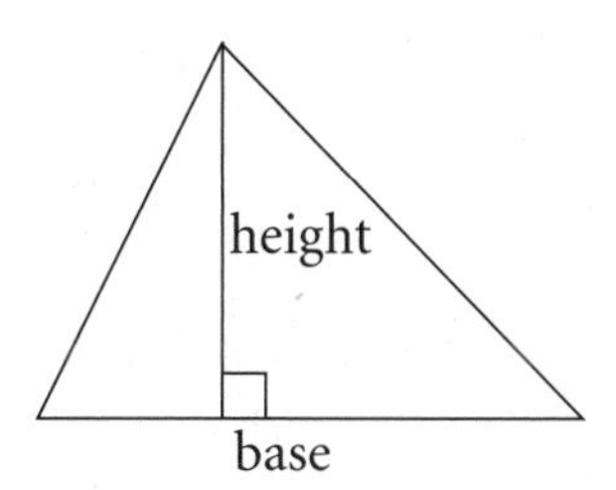

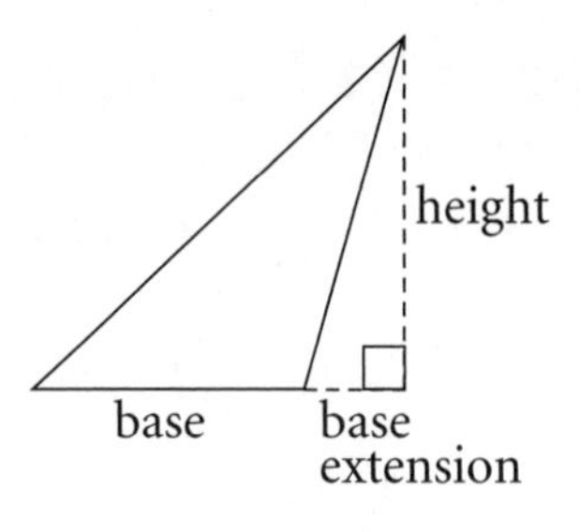

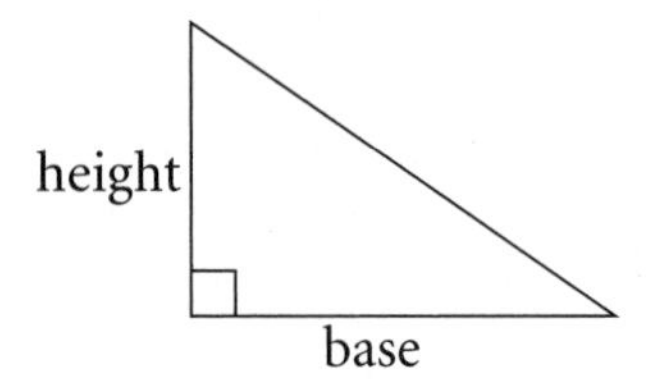

Hook

Think of a triangle as being **half** a rectangle. The area of that triangle (as well as the area of the largest triangle that fits inside a rectangle) is **half** the area of a rectangle whose sides are the base and height of the triangle.

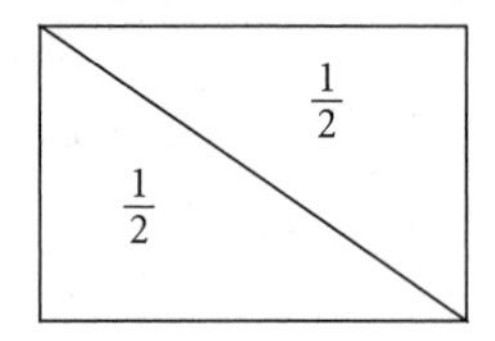

Example

Find the area of a triangle with a 2-inch base and a 3-inch height.

1. Draw the triangle as close to scale as you can.

2. Label the size of the base and height.

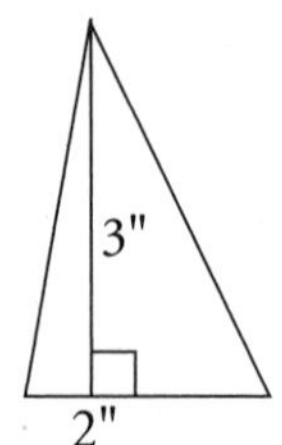

3. Write the area formula; then substitute the base and height numbers into it: $area = \frac{1}{2}$ (base × height)

   $area = \frac{1}{2}(2 \times 3) = \frac{1}{2} \times 6$

4. The area of the triangle is **3 square inches.** $area = \mathbf{3}$

Practice

_____ **11.** The base of a triangle is 14 feet long, and the height is 5 feet. How many square feet of area is the triangle?

_____ **12.** If a triangle with 100 square feet of area has a base that is 20 feet long, how tall is the triangle?

Find the area of the following triangles.

_____ **13.**

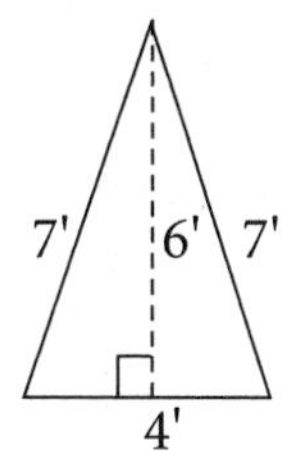

_____ **14.**

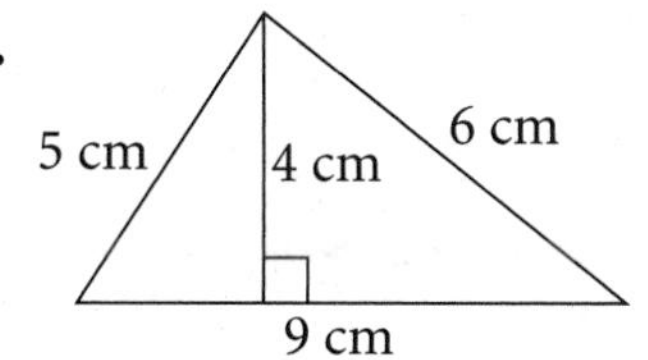

_____ **15.**

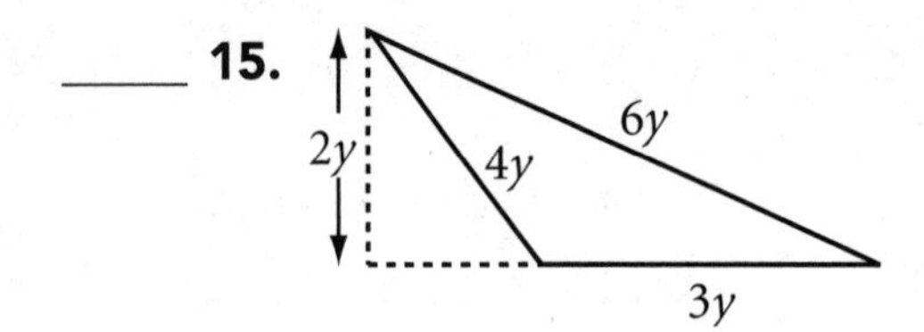

_____ **16.**

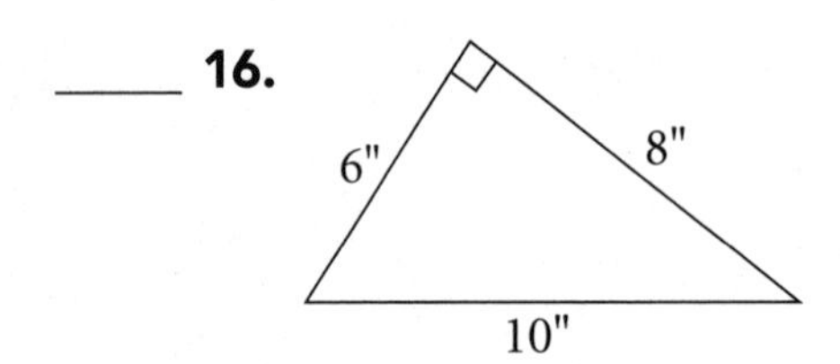

_____ **17.**

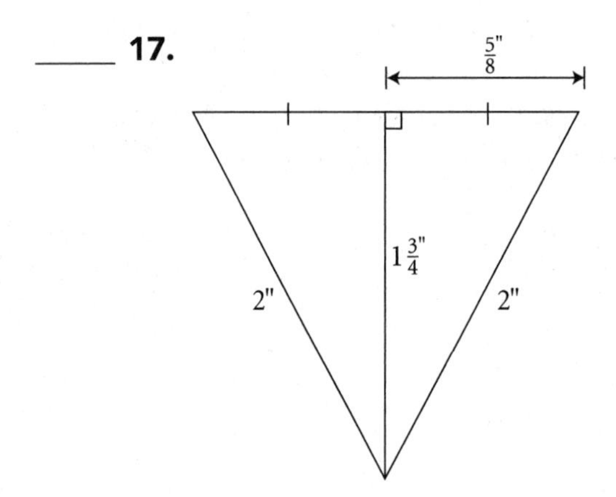

_____ **18.**

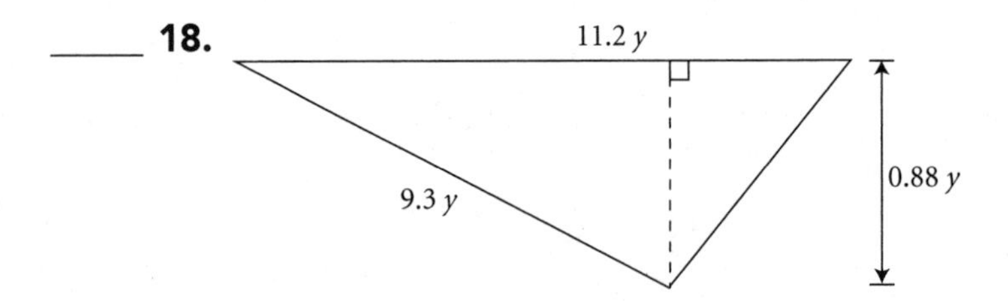

19.

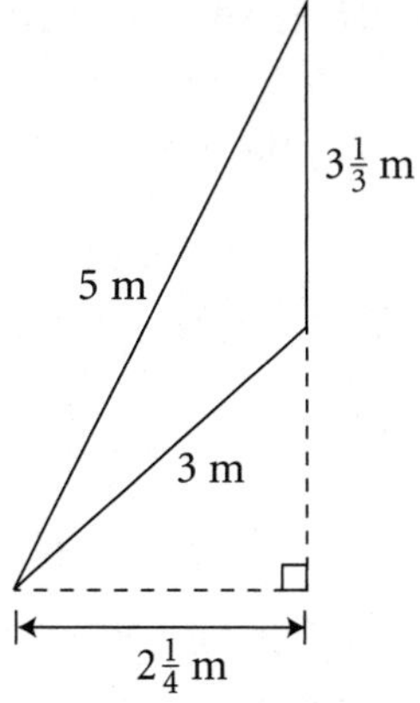

Area of Circles

Even though it's hard to imagine how many squares it would take to fill the space that a circle occupies, that is what we are doing when we find the area of a circle. Use the following illustration to estimate how many squares are needed to completely fill in a circle with a radius of 2 units:

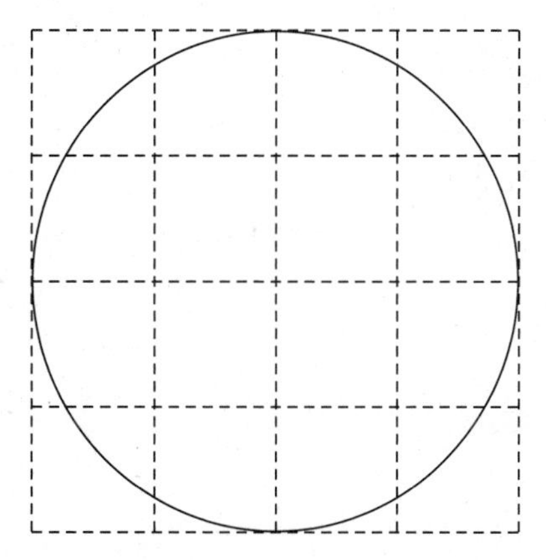

If your estimate is between 12 and 13, then you did a great job! Now that you have a visual idea of what the area of a circle represents, let's look at how to solve for the area of a circle by using the formula.

To determine the area of a circle, use this formula:

$$\textit{Area} = \pi r^2$$

The formula can be written out as $\pi \times r \times r$.

Hook

To avoid confusing the area and circumference formulas, just remember that *area* is always measured in *square* units, as in 12 *square yards* of carpeting. This will help you remember that the area formula is the one with the *squared* term in it.

Example

Find the area of the following circle, rounded to the nearest tenth.

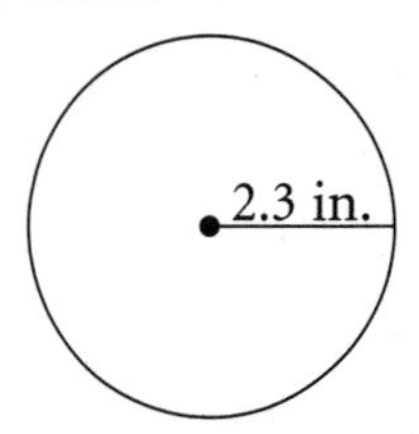

1. Write the area formula: $A = \pi r^2$
2. Substitute 2.3 for the radius: $A = \pi \times 2.3 \times 2.3$
3. On a multiple-choice test, look at the answer choices to determine whether to use π or the *value of* π (decimal or fraction) in the formula. If the answers don't include π, use 3.14 for π (because the radius is a decimal):

   $A = 3.14 \times 2.3 \times 2.3$

   $\mathbf{A = 16.6}$

   If the answers include π, multiply and round:

   $A = \pi \times 2.3 \times 2.3$

   $\mathbf{A = 5.3\pi}$

Both answers—**16.6 square inches** and **5.3π square inches**—are considered correct.

Example

What is the diameter of a circle whose area is 9π square centimeters?

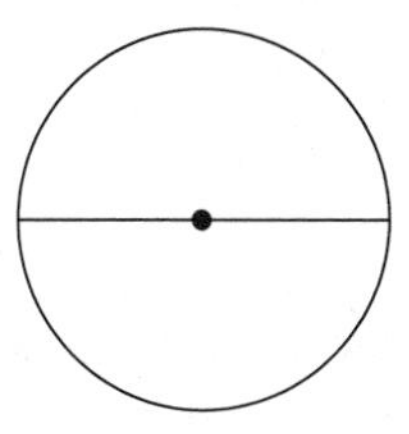

1. Draw a circle with its diameter (to help you remember that the question asks for the diameter); then write the area formula.

   $A = \pi r^2$

2. Substitute 9π for the area and solve the equation:

   $9\pi = \pi r^2$

   $9 = r^2$

   Since the radius is 3 centimeters, the diameter is **6 centimeters.**

   $3 = r$

CAUTION

r^2 does not mean $2 \times r$. It means $r \times r$.

Practice

Find the area.

_____ **20.**

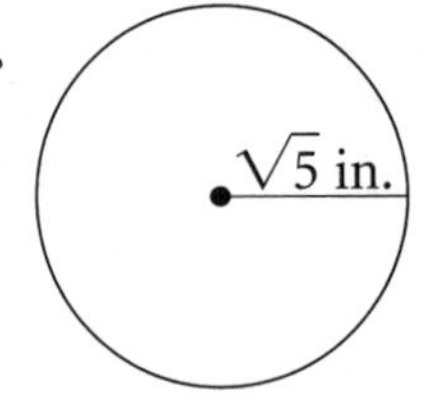

_____ **21.**

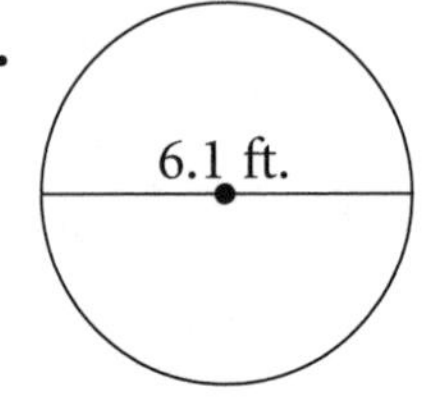

Find the radius and diameter.

_____ **22.**

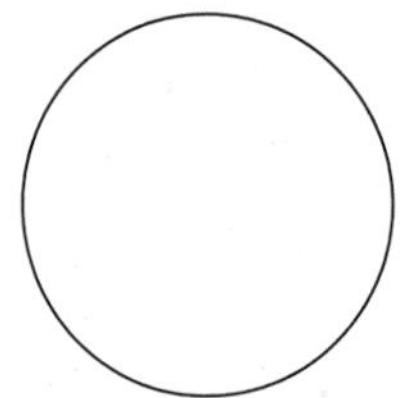

Area = 49π ft.^{2}

_____ **23.**

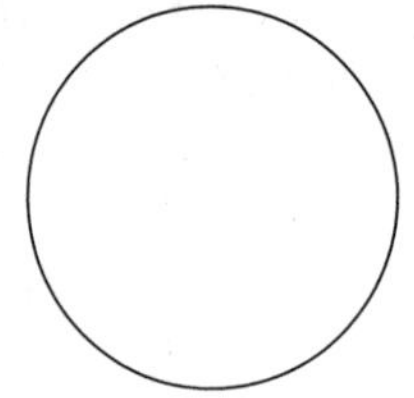

Area = πm^2

Word Problems

_____ **24.** What is the area in square inches of the bottom of a jar with a diameter of 6 inches?

_____ **25.** James Band is believed to be hiding within a 5-mile radius of his home. What is the approximate area, in square miles, of the region in which he may be hiding? If Lieutenant Reaves needs to dispatch one search hound for every 3 square miles of territory, how many search hounds should he request to look for James Band?

_____ **26.** Approximately how many more square inches of pizza are in a 12-inch diameter round pizza than in a 10-inch diameter round pizza? Next determine the percentage increase of area when going from a 10-inch pizza to a 12-inch pizza.

_____ **27.** The following figure represents the placement of a hot tub with a radius of 1 meter on a square wooden deck with a side length of 5 meters. A can of deck varnish is enough to cover an area of 5 square meters. Find the area of the wooden deck (the shaded region) and determine how many cans of varnish need to be purchased in order to fully cover the deck.

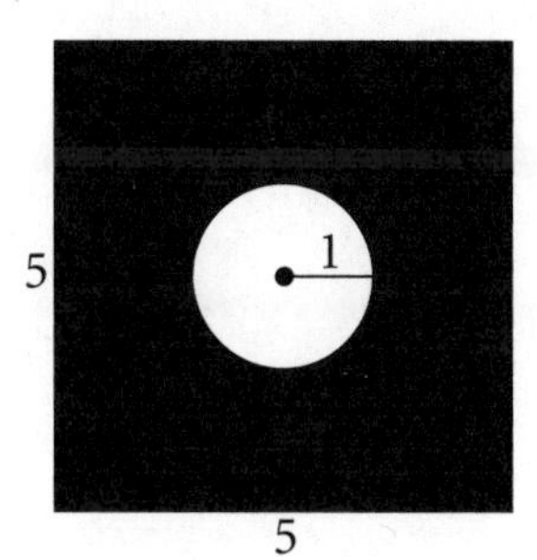

_____ **28.** Farmer McDonald knows that the circumference of her silo is 75.4 feet. How many square feet of hay will she need to fully cover the ground within the silo if she would like to use it as a pigpen during several stormy months?

TRY THIS

When you are at home, take a look at your kitchen and make a rough estimate of how many 1-foot by 1-foot tiles would be needed to cover your kitchen floor. Then make a sketch of your kitchen floor and use a tape measure to find the dimensions of your kitchen and then label your drawing. Next, calculate the area of your kitchen floor. This will represent how many 1-foot square tiles would be needed to cover it. How close was your rough estimate? If your floor already has 1-foot square tiles, count the tiles and compare with your answer. How much would it cost to retile your kitchen floor if square foot tiles cost $4 each?

Answers

1. 17 in.^{2}
2. 6.25 mm^2
3. 4.5 feet
4. 5 yards
5. 4 cans
6. 99 tiles
7. Area of yard = 4,600 ft.^{2}, cost for artificial grass = $16,100
8. A coop of 4 ft. by 4 ft. would give the most space for the chickens to walk around: 16 ft.^{2}
9. If given a fixed perimeter to create a coop, a square gives the most area. A long and skinny rectangle gives the smallest area.
10. Since a square gives the most area for a fixed perimeter, it would be best to build a square house, and not a rectangular house. This would keep the construction costs down while keeping the living space as large as possible.
11. 35 square feet
12. 10 feet
13. 12 feet^2
14. 18 cm^2
15. 3 y^2
16. 24 in.^{2}
17. $\frac{35}{64}$ in.^{2}
18. 4.928 units^2
19. $3\frac{3}{4}$ m^2
20. 5π in.^{2} or 15.7 in.^{2}
21. 9.3π in.2 or 29.2 in.^{2}
22. $r = 7$ ft.; $d = 14$ ft.
23. $r = 1$ m; $d = 2$ m
24. 9π in.^{2} or 28.26 in.^{2}
25. 78.5 mi.^{2}; 26 search hounds

26. 35 in.^{2}; $\frac{35}{79} \times 100 = 44\%$ larger

27. 21.86 m^2; $\frac{21.86}{5} = 4.4$ cans, so you will need to purchase 5 cans since 4 cans will not be enough.

28. 452 square feet of hay

CHAPTER 17 Volume

"The mathematical sciences particularly exhibit order, symmetry, and limitation; and these are the greatest forms of the beautiful."

—ARISTOTLE, Greek philosopher (384 B.C.E.–322 B.C.E.)

CHAPTER SUMMARY

In this chapter you will learn about volume and how to calculate the volume for a variety of three-dimensional figures.

"How much volume would you like?" This really should be the question that you are asked when ordering a glass of juice in a restaurant. It is not really the *size* of the glass that you are interested in, but the *amount* of juice that goes into that glass. **Volume** refers to the amount of space occupied *inside* a three-dimensional object. **Volume** is an important concept to consider when talking about three-dimensional figures such as shipping containers, fish tanks, and rooms in an office building. Similarly to the way area is the measure of how many square units cover a flat space, volume is the measure of how many cube units *fill up* a three-dimensional space. A unit cube is a three-dimensional box that has equal length, width, and height. Think about neatly arranging one-inch cubes, side-by-side, in a small shoebox. The total number of cubes that it would take to fill the box completely would be the volume of the box in cubic inches.

> ***Volume* is the amount of space enclosed by a three-dimensional object.**

Perimeter, area, and volume are three distinct measurements that are each important for different applications. The *perimeter* of a garden box will determine the length of wood a carpenter must purchase in order create the frame. A landscape architect needs the *area* of the garden box in order to determine the square footage of anti-weed cloth that will line the bottom of the box. The *volume* of this finished garden box will show the gardener how many cubic feet of potting soil must be purchased to fill the box completely.

Practice

For questions 1 through 6, state whether perimeter, area, or volume would be used for each situation.

_____ **1.** Max needs to paint the outside of a box black so that it can be used as a prop on stage for a play.

_____ **2.** Rikki needs to buy the right-sized air conditioner to cool down her studio apartment.

_____ **3.** Cole was late for soccer practice and has to run 5 laps around the field.

_____ **4.** Caitlin wonders how much it will cost to fill the pool with water if each gallon from her hose costs $0.07.

_____ **5.** Stefanus is buying fabric to make a tablecloth.

_____ **6.** Daniel wants to put a tile border around the bay window in his living room.

Finding the Volume of Rectangular Prisms

A rectangular prism is a three-dimensional shape that has a rectangle as its face. Most boxes and rooms are rectangular prisms. To understand how volume is calculated, we will consider the rectangular prism in the following figure:

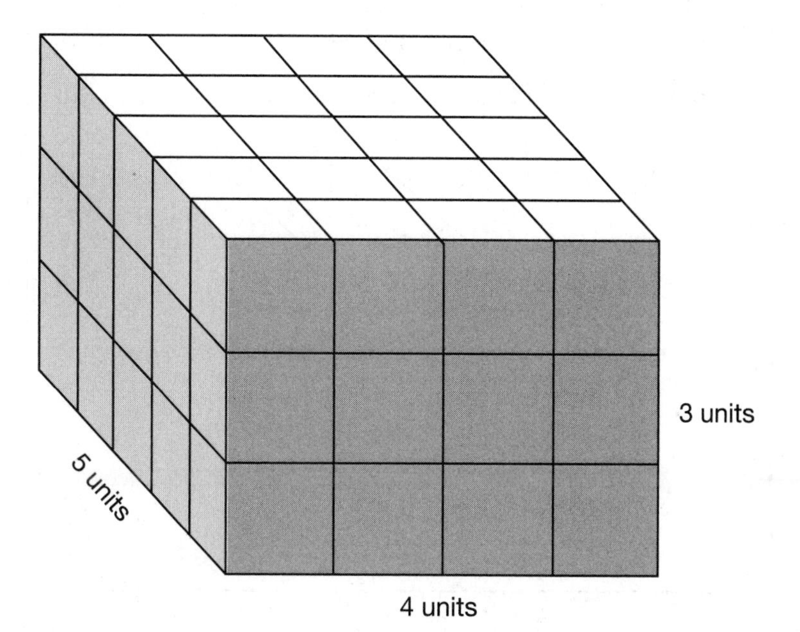

The face of the rectangular prism, which is shaded the darkest, has a length of 4 units and a height of 3 units. Therefore the area of the front face is $4 \times 3 = 12$ square units. The width of this rectangular prism is 5 units. Can you notice that for each of the 12 cubes on the front face there is a row of 5 cubes extending toward the back of the prism? Knowing this, you can find the volume by multiplying the number squares on the front face (12) by the number of cubes behind it, which is represented as the width (5). Therefore, the number of cubes that make up this prism is $12 \times 5 = 60$. The volume of the prism is therefore 60 cubic units, which is written 60 units^3. The **volume** of this **rectangular prism** was calculated by multiplying the length, width, and height. Volume is always expressed with an exponent of 3 after the units, to represent the three different dimensions volume measures and to remind us that the volume of any figure is comprised of cubes.

volume of rectangular prism $= \boldsymbol{l \times w \times h}$

Sample Question 1

Find the volume of the rectangular prism:

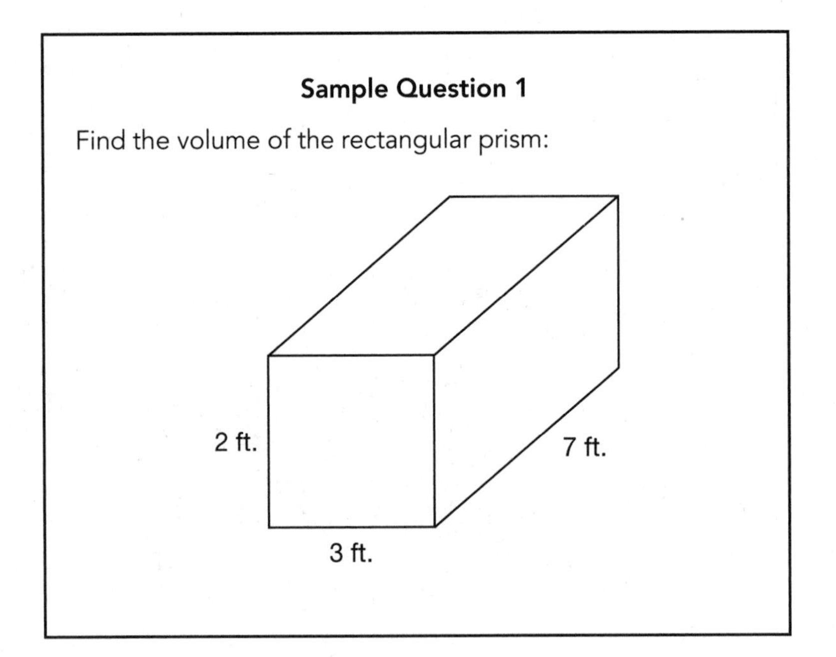

Finding the Volume of Cubes

A cube is a special type of rectangular prism where the length, width, and height are all the same length. Similarly to how the side length of squares is named *s*, the side lengths of a cube are also referred to as *s*. The formula for finding the volume of a cube is the product of all three sides:

$$\textbf{volume of a cube} = \textbf{side} \times \textbf{side} \times \textbf{side}$$

(or)

$$\textbf{volume of a cube} = s^3$$

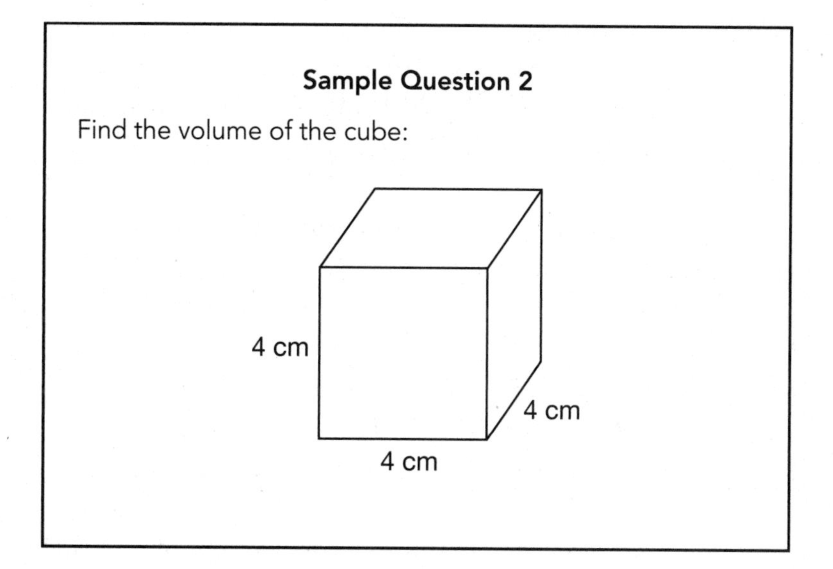

Practice

_____ **7.** Luke built a garden box that is 6 feet long by 3 feet wide by 2 feet tall. How many cubic feet of soil will he need to buy to fill it?

_____ **8.** What has a larger volume? A square box with a side length of 2 feet, or a rectangular box that is 5 feet long by 3 feet wide by $\frac{1}{2}$ foot tall? Find the volume of these two prisms to determine your answer.

_____ **9.** Willa is a contractor who is ordering the air-conditioning units for the living rooms in a new apartment building. There is a Veteran's Day sale on an air conditioner that cools rooms up to 1,500 cubic feet, but she's not sure whether that will be powerful enough for the large living rooms in each apartment. Willa remembers that she had to order 225 square feet of carpet for each living room and knows that the ceilings are 9 feet high. Calculate the volume of each living room to see whether the on-sale air conditioners are powerful enough for them.

_____ **10.** True or false: If the side length of a cube is doubled, then the volume of the new cube is also doubled. Create an example to support your answer.

_____ **11.** Evie works for a company that keeps clients' pools clean and healthy by maintaining the proper chemical balance. One of the products she adds to pools each week calls for 1 tablespoon for every 100 cubic feet of water. If Gray's pool is 15 feet long by 12 feet wide by 4 feet deep, how many cubic feet of water does it contain? How many tablespoons of chemical will Evie need?

Finding the Volume of Triangular Prisms and Cylinders

A **right prism** is a prism with two congruent polygon bases that are perpendicular to its sides. A fresh stick of butter is a right prism, but if you cut it on an angle, then it is no longer a right prism because its face will not be perpendicular to its sides. Given *any* right prism, if you can calculate the area of one of its congruent bases, then multiplying that area by the height of the prism will give you the volume. To understand this, look at the following figure:

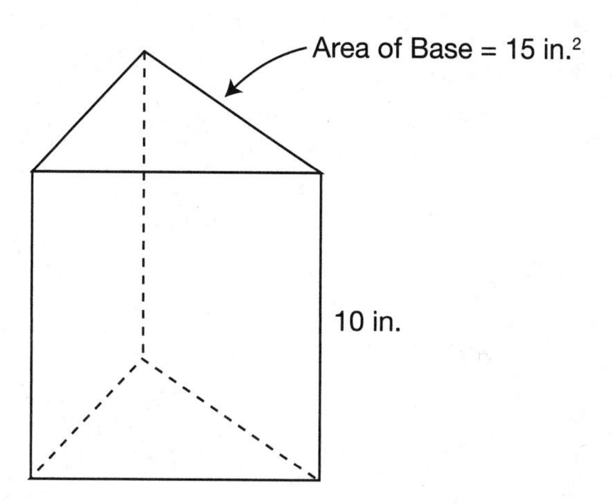

Since it is given that the area of the top triangular face is 15 in.^{2}, then multiplying that area by the 10-inch height will yield the volume of 150 inches^3. As a formula, this technique looks like this:

volume of a right prism = area of base × height

A **cylinder** is not a prism since its bases are circles and not polygons, but this same technique can be applied to finding the volume of that glass of orange juice we began this chapter with. As long as you can calculate the area of the circular base, that area can be multiplied by the height in order to find the volume.

volume of a cylinder = area of circular base × height

Sample Questions 3 and 4

3. The cylinder in figure A represents the dimensions of a pool in feet. Find its volume.

4. Find the volume of the prism in figure B.

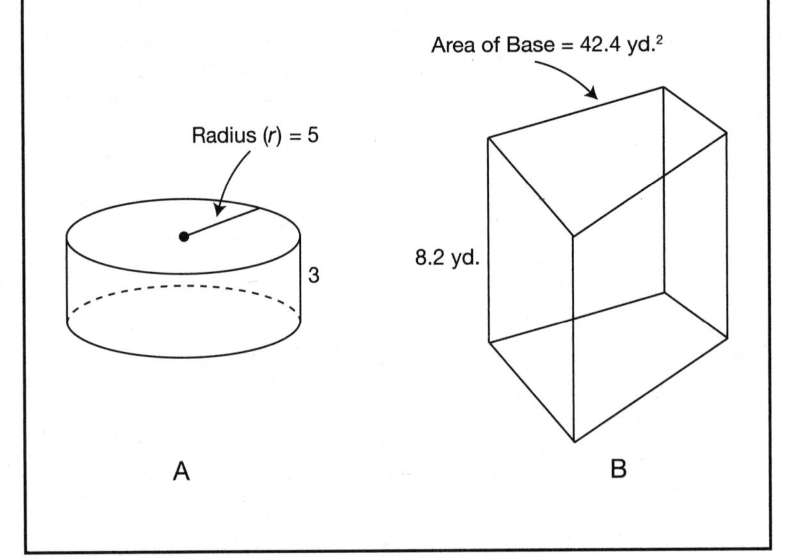

Special formulas are used find the volumes of pyramids, cones, and spheres, but if you need to calculate those on the job, it is probably best to look those formulas up. For now, test your understanding of volume in right prisms and cylinders.

Practice

_____ **12.**

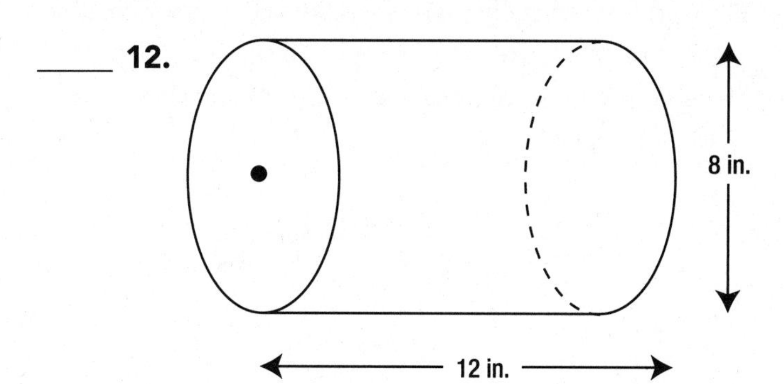

_____ **13.**

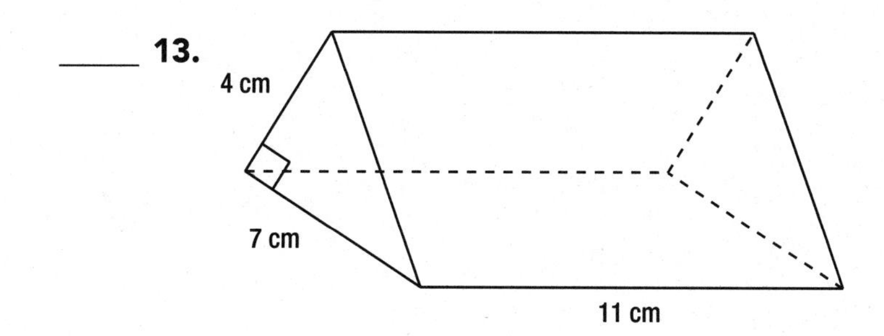

_____ **14.** Can the formulas presented in this chapter be applied to this organically shaped vase? It has a curvy base, but the bases are at right angles to the sides. If the volume can be determined, find it. If it can't be calculated, explain why.

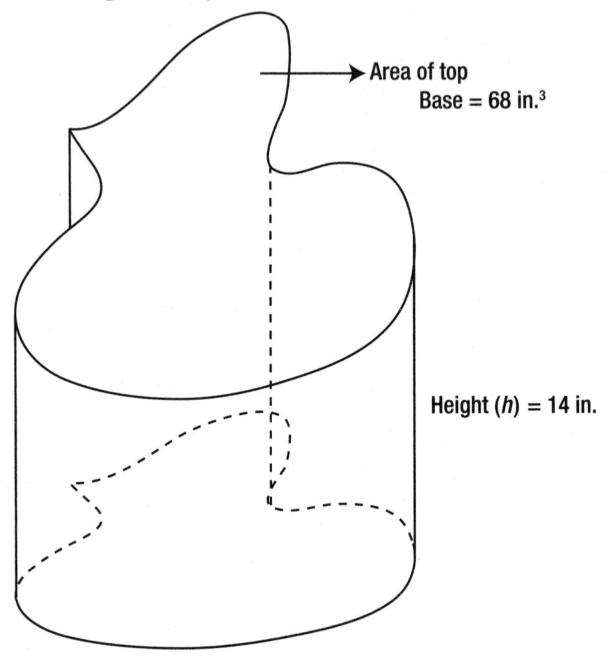

_____ **15.** A huge cylindrical fish tank sits in the Casa Del Mar hotel lobby. It is 3 feet wide by 12 feet tall. What is the volume in cubic feet of this custom tank? Use 3.14 for pi. If one cubic foot is equivalent to about 7.5 gallons, how many gallons of water does the fish tank hold?

_____ **16.** Toblerone makes chocolate bars in the shape of triangular prisms. One of their most popular sizes has a triangular face that is 20mm wide and 26mm tall. This chocolate bar is 200mm long. What is the volume of chocolate in this bar?

_____ **17.** Boulder Mountain Chocolate makes chocolate bars as rectangular prisms. Their best-selling size is 180mm long by 30mm wide and 9mm tall. What is the volume of this chocolate bar and how does it compare in size to the Toblerone bar from question 16?

_____ **18.** Mr. Ming is building an art studio in his backyard that has the dimensions given in the following illustration. He would like the studio to have a vaulted ceiling of the exposed roof, so that it is as open as possible. However, his daughter argues that doing so would increase his heating bills in the winter and air conditioning bills in the summer by more than 20% because of the additional volume that will need to be heated and cooled. She thinks he should install a ceiling at 12' instead of leaving the room open to the roof. Calculate the volumes of the studio with and without an open vaulted ceiling. Then use your knowledge of percentage increase to calculate the percentage increase in volume that an open ceiling would create.

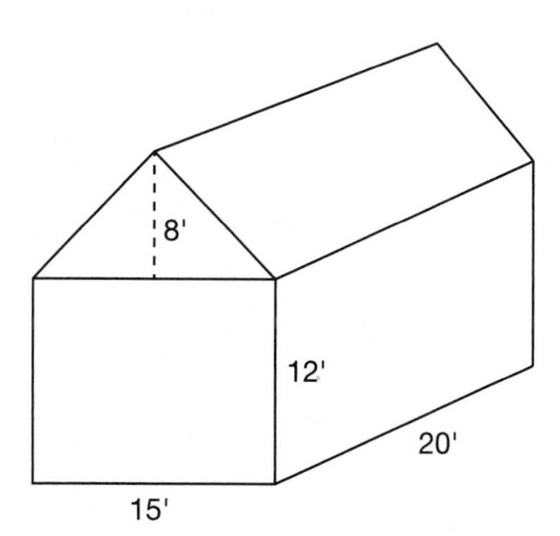

TRY THIS

Take some measurements of your bathtub and estimate how many cubic feet of water it takes to fill it up for a bath. Now figure out how many gallons it uses to take a bath, using the fact that 1 cubic foot contains approximately 7.5 gallons of water. If a low-flow showerhead uses about 1.5 gallons of water per minute, how many minutes would you need to shower in order to use as much water as a single bath takes?

Answers

Practice Problems

1. Area (Max will have to add the areas of all 6 sides of the box.)
2. Volume
3. Perimeter
4. Volume
5. Area
6. Area
7. 36 cubic feet of soil
8. Volume of square box = 8 ft.^{3} and volume of rectangular box = 7.5 ft.^{3}, so the square box has a bigger volume.
9. Volume = 2,025 ft.^{3} and on-sale air conditioners will be too small to cool these rooms.
10. False! If the side length of a cube is doubled, then the volume of the new cube is not just doubled. Example: The volume of a cube with $s = 5$ is 125 units^3, but the volume of a cube with $s = 10$ is 1,000 units^3. (It increases by 8 times.)
11. Volume = 720 cubic feet so Evie will need 7 tablespoons of this particular product for Gray's pool.
12. 192π or 602.88 in.^{3}
13. 154 cm^3
14. Since the area of the base is given and the sides are parallel, the volume can be found by multiplying the area by the height. Volume = 952 in.^{3}
15. Volume = 84.78 ft.^{3}, which is approximately 635.85 gallons of water.
16. Volume = 52,000 mm^3
17. Volume = 48,600 mm^3. It is slightly smaller than the Toblerone bar, but close in size.

18. Volume of rectangular prism = 3,600 ft.^{3}
Volume of triangular prism roof = 1,200 ft.^{3}
Volume of studio with open roof = 4,800 ft.^{3}
Percentage increase in volume going from 12' ceiling to open vaulted ceiling: 33% increase in volume from 3,600 ft.^{3} to 4,800 ft.^{3}

Sample Question 1

1. Volume of a rectangular prism = $l \times w \times h$

2. Volume = $2 \times 3 \times 7$

3. Volume = 42 feet^3 or 42 cubic feet

Sample Question 2

1. Volume of a cube = s^3

2. Volume = 4^3

3. Volume = 64 feet^3 or 64 cubic feet

Sample Question 3

1. Volume of a cylinder = area of circular base × height

2. Volume = $\pi r^2 \times$ height

3. Volume = $\pi 5^2 \times 3$

4. Volume = 75π cubic feet or 235.5 ft^3

Sample Question 4

1. Volume of a right prism = area of base × height

2. Volume = 42.4×8.2

3. Volume = 347.68 yd^3

CHAPTER

18 Miscellaneous Math

Arithmetic is one of the oldest branches, perhaps the very oldest branch, of human knowledge; and yet some of its most abstruse secrets lie close to its tritest truths.

—Norman Lockyer, English scientist (1836–1920)

CHAPTER SUMMARY

This chapter contains miscellaneous math topics that don't fall into the other lessons. Achieving a comfort level with these tidbits will support your success in other areas, such as word problems.

This lesson covers a variety of math topics that often appear in real-life situations:

- Positive and negative numbers
- Order of mathematical operations
- Working with length unit conversions

- Squares and square roots
- Solving algebraic equations

Positive and Negative Numbers

Negative numbers abound in everyday life. Stocks drop in price, temperatures dip below zero, and interest fees turn up on credit card statements—these are just some examples of how negative numbers are used in applied problems. Positive and negative numbers, also called signed numbers, can be visualized as points along the number line:

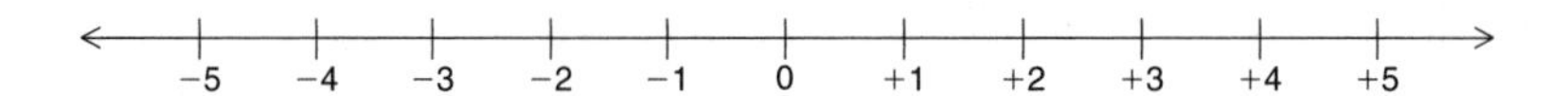

Numbers to the left of 0 are *negative* and those to the right are *positive.* Zero is neither negative nor positive. If a number is written without a sign, it is assumed to be *positive.* On the negative side of the number line, numbers further to the left are actually smaller. For example, –5 is *less than* –2. Therefore the *larger* a negative number is, the *smaller* its value is. For example, –28 degrees is a lower temperature than –10 degrees.

Arithmetic with Positive and Negative Numbers

The following table illustrates the rules for doing arithmetic with signed numbers. Notice that when a negative number follows an operation (as it does in the second example), it is enclosed in parentheses to avoid confusion.

RULE	EXAMPLE
ADDITION	
If both numbers have the same sign, just add them. The answer has the same sign as the numbers being added.	$3 + 5 = 8$ $-3 + (-5) = -8$
If both numbers have different signs, subtract the smaller number from the larger. The answer has the same sign as the one whose numeral is larger.	$-3 + 5 = 2$ $3 + (-5) = -2$
If both numbers are the same but have opposite signs, the sum is zero.	$3 + (-3) = 0$
SUBTRACTION	
To subtract one number from another, change the sign of the number to be subtracted, and then add as above. $3 - 5 =$ $3 + (-5) = -2$ $-3 - 5 =$ $-3 + (-5) = -8$ $-3 - (-5) =$ $-3 + 5 = 2$	

TIP

Sometimes subtracting with negatives can be tricky. Remembering "keep-switch-switch" can be a helpful way to recall that you should **keep** the first sign the same, **switch** the minus to a plus, and **switch** the sign of the third term.

Examples: $-5 - 4$ would become $-5 + -4$, and $27 - (-9)$ would become $27 + 9$

RULE	EXAMPLE
MULTIPLICATION	
Multiply the numbers together. If both numbers have the same sign, the answer is positive; otherwise, it is negative. If at least one number is zero, the answer is zero.	$3 \times 5 = 15$ $-3 \times (-5) = 15$ $-3 \times 5 = -15$ $3 \times (-5) = -15$ $3 \times 0 = 0$
DIVISION	
Divide the numbers. If both numbers have the same sign, the answer is positive; otherwise, it is negative. If the top number is zero, the answer is zero.	$15 \div 3 = 5$ $-15 \div (-3) = 5$ $15 \div (-3) = -5$ $-15 \div 3 = -5$ $0 \div 3 = 0$

CAUTION

You cannot divide by zero.

TIP

To help remember the sign rules of multiplication, think of the following: Let *being on time/starting on time* be metaphors for a positive number and *being late/starting late* be metaphors for a negative number.

Being on time to something that starts on time is a good thing. $(+ \times + = +)$

Being late to something that starts on time is a bad thing. $(- \times + = -)$

Being on time to something that starts late is a bad thing (because you'll have to wait around). $(+ \times - = -)$

Being late to something that starts late is a good thing (because now you're on time!). $(- \times - = +)$

Practice

Use the rules in the previous tables to solve the following problems with signed numbers.

_____ **1.** $20 + (-50)$

_____ **2.** $-12 + (-16)$

_____ **3.** $-7 - (-10)$

_____ **4.** $8 - (-25)$

_____ **5.** -12×-5

_____ **6.** $-24 \div 6$

For questions 7 through 11, use signed integers to write equations that represent the following situations and then solve them.

_____ **7.** The temperature is –9° Fahrenheit and it drops another 6°. What is the current temperature?

_____ **8.** A company goes out of business and a debt of $10,000 is now shared by 5 people. How much must each person pay to clear the debt?

_____ **9.** Chelsea's checking account has a balance of $40. She buys a $120 jacket with her debit card, which the bank pays since she has overdraft protection. However, the bank then charges Chelsea's account a $35 overdraft fee. What is her current balance in the account?

_____ **10.** VJ buys three shirts that cost $14 each, 2 pairs of pants that cost $32 each, and 6 pairs of socks that cost $2.50 each. He pays with a $100 bill and a $50 bill. How much change will he get?

_____ **11.** Rachael has a credit card bill of $1,200 that accrues interest at a rate of 1.5% per month. If she does not make any payment this month, what is the interest that will be added to her balance?

Order of Operations

When an expression contains more than one operation—like $2 + 3 \times 4$—you need to know the order in which to perform the operations.

Notice that the order in which you do the operations changes the result:

1. *Addition first:* $2 + 3 \times 4 = 5 \times 4 = \underline{20}$
2. *Multiplication first:* $2 + 3 \times 4 = 2 + 12 = \underline{14}$

Only one of these answers, 20 or 14, can be correct—which one is it? To answer this question, let's look at the correct order in which all arithmetic equations must be solved:

1. *Parentheses:* Evaluate everything inside parentheses before doing anything else.
2. *Exponents:* Next, evaluate all exponents.
3. *Multiplication and Division:* Go from left to right, performing each multiplication and division as you come to it.

4. *Addition and Subtraction:* Go from left to right, performing each addition and subtraction as you come to it.

Hook

The following sentence is commonly used to remember this order of operations: **P**lease **E**xcuse **M**y **D**ear **A**unt **S**ally.

Practice

Use the order of operations to solve these problems.

_____ **12.** $3 + 6 \times 2 = ?$

_____ **13.** $4 \times 2 + 3 = ?$

_____ **14.** $(3 + 5) \div 2 + 2 = ?$

_____ **15.** $2 + 5 \times 6 - 4 + 7 = ?$

_____ **16.** $2 \div 5 \times 3 - 1 = ?$

_____ **17.** $2 \div 4 + 3 \times 4 \div 8 = ?$

_____ **18.** $(5 \div \frac{1}{5}) - 25 \times 0.5 = ?$

Working with Length Units

The United States uses the *English system* to measure length; however, Canada and most other countries in the world use the *metric system* to measure length. Using the English system requires knowing many different equivalences, but you're probably used to dealing with these equivalences on a daily basis. Mathematically, however, it's simpler to work in metric units because their equivalences are all multiples of 10. The meter is the basic unit of length, with all other length units defined in terms of the meter.

ENGLISH SYSTEM	
UNIT	**EQUIVALENCE**
foot (ft.)	1 ft. = 12 in.
yard (yd.)	1 yd. = 3 ft. 1 yd. = 36 in.
mile (mi.)	1 mi. = 5,280 ft. 1 mi. = 1,760 yds.

METRIC SYSTEM	
UNIT	**EQUIVALENCE**
meter (m)	Basic unit A giant step is about 1 meter long.
centimeter (cm)	100 cm = 1 m Your index finger is about 1 cm wide.
millimeter (mm)	10 mm = 1 cm; 1,000 mm = 1 m Your fingernail is about 1 mm thick.
kilometer (km)	1 km = 1,000 m Five city blocks are about 1 km long.

Length Conversions

Many jobs require the ability to perform length conversions within a particular system. One example is when the measurements of a plot of land are originally taken in feet, but then need to be converted into miles. An easy way to convert from one unit of measurement to another is to multiply by an ***equivalence ratio***. Such ratios don't change the value of the unit of measurement because each ratio is equivalent to 1. The equivalence ratio should have the unit you are converting *from* in the denominator and the unit you are converting *to* in the numerator. For example, if you want to convert feet to miles, you would use the equivalence ratio $\frac{\text{1 mile}}{\text{5,280 feet}}$. The following boxes contain the majority of equivalence ratios you will need to use. Notice that in the following two tables, all the ratios have values in the numerator and denominator that are equivalent, so each ratio has a value of 1.

ENGLISH SYSTEM	
TO CONVERT BETWEEN	**MULTIPLY BY THIS RATIO**
inches and feet	$\frac{12\text{ in.}}{1\text{ ft.}}$ or $\frac{1\text{ ft.}}{12\text{ in.}}$
inches and yards	$\frac{36\text{ in.}}{1\text{ yd.}}$ or $\frac{1\text{ yd.}}{36\text{ in.}}$
feet and yards	$\frac{3\text{ ft.}}{1\text{ yd.}}$ or $\frac{1\text{ yd.}}{3\text{ ft.}}$
feet and miles	$\frac{5{,}280\text{ ft.}}{1\text{ mi.}}$ or $\frac{1\text{ mi.}}{5{,}280\text{ ft.}}$
yards and miles	$\frac{1{,}760\text{ yds.}}{1\text{ mi.}}$ or $\frac{1\text{ mi.}}{1{,}760\text{ yds.}}$

METRIC SYSTEM	
TO CONVERT BETWEEN	**MULTIPLY BY THIS RATIO**
meters and millimeters	$\frac{1{,}000\text{ mm}}{1\text{ m}}$ or $\frac{1\text{ m}}{1{,}000\text{ mm}}$
meters and centimeters	$\frac{100\text{ cm}}{1\text{ m}}$ or $\frac{1\text{ m}}{100\text{ cm}}$
meters and kilometers	$\frac{1{,}000\text{ m}}{1\text{ km}}$ or $\frac{1\text{ km}}{1{,}000\text{ m}}$
millimeters and centimeters	$\frac{10\text{ mm}}{1\text{ cm}}$ or $\frac{1\text{ cm}}{10\text{ mm}}$

Example

Convert 7 yards to feet.

Since you are converting from yards to feet, use an equivalence ratio that has yards in the denominator and feet in the numerator. That ratio will be $\frac{3\text{ ft.}}{1\text{ yd.}}$ since 3 feet are in 1 yard. Notice that we use $\frac{3\text{ ft.}}{1\text{ yd.}}$ rather than $\frac{1\text{ yd.}}{3\text{ ft.}}$ because the *yards* cancel during the multiplication. Multiply 7 yards by $\frac{3\text{ ft.}}{1\text{ yd.}}$:

$$7\text{ yds.} \times \frac{3\text{ ft.}}{1\text{ yd.}} = \frac{7\,\cancel{\text{yds.}} \times 3\text{ ft.}}{1\,\cancel{\text{yd.}}} = \mathbf{21\text{ ft.}}$$

Example

Convert 31 inches to feet *and inches.*

First, multiply 31 inches by the ratio $\frac{1\text{ ft.}}{12\text{ in.}}$:

$$31\text{ in.} \times \frac{1\text{ ft.}}{12\text{ in.}} = \frac{31\,\cancel{\text{in.}} \times 1\text{ ft.}}{12\,\cancel{\text{in.}}} = \frac{31}{12}\text{ ft.} = 2\frac{7}{12}\text{ ft.}$$

Then, change the $\frac{7}{12}$ portion of $2\frac{7}{12}$ ft. to inches:

$$\frac{7\text{ ft.}}{12} \times \frac{13\text{ in.}}{1\text{ ft.}} = \frac{7\,\cancel{\text{ft.}} \times \cancel{12}\text{ in.}}{\cancel{12} \times 1\text{ ft.}} = 7\text{ in.}$$

Thus, 31 inches is equivalent to both **$2\frac{7}{12}$ ft.** and **2 feet 7 inches.**

TIP

Time Conversions: Word problems love giving information in mixed units of time. Since **5 minutes 42 seconds ≠ 5.42** minutes, it's always best to convert mixed information into the smallest unit. Change 5 minutes 42 seconds into seconds by multiplying the minutes by 60 and adding on the seconds: $5 \times 60 + 42 = 342$ seconds. You can change your answer back to minutes and seconds later by dividing it by 60. The whole number part of your answer will be the minutes, and the remainder will be the number of seconds. Example: 343 seconds: $343 \div 60 = 5$, remainder 43, which is 5 minutes 43 seconds. (Use the same tips for working with minutes and hours.)

Practice

Convert each of the following units to the desired units.

19. 4.75 ft. = ______ in.

20. 50 in. = ______ ft. ______ in.

21. 3.5 yd. = _____ in.

22. 117 in. = _____ yd. _____ in.

23. 3.2 mi. = _____ ft.

24. 6,336 ft. = _____ mi.

25. 16.8 m = _____ cm

26. 85.62 km = _____ m

27. 22 in. = _____ ft.; 22 in. = _____ ft. _____ in.

28. $4\frac{1}{2}$ ft. = _____ yds.;
$4\frac{1}{2}$ ft. = _____ yds. _____ ft. _____ in.

29. 7,920 ft. = _____ mi.;
7,920 ft. = _____ mi. _____ ft.

30. 1,100 yds. = _____ mi.

31. 342 mm = _____ cm;
342 mm = _____ cm _____ mm

32. 165 mm = _____ km

Addition and Subtraction with Length Units

Finding the perimeter of a figure may require adding lengths of different units.

Example

Find the perimeter of the following figure.

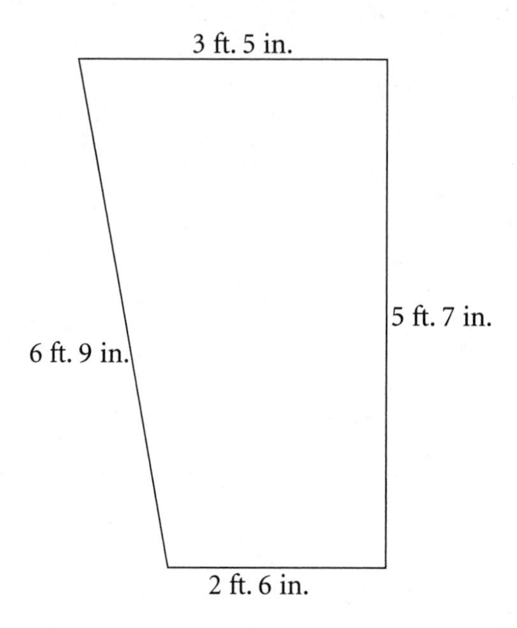

To add the lengths, add each column of length units separately:

5 ft.	7 in.
2 ft.	6 in.
6 ft.	9 in.
+ 3 ft.	5 in.
16 ft.	**27 in.**

Since 27 inches is more than 1 foot, the total of **16 ft. 27 in.** must be simplified:

- Convert 27 inches to feet and inches:
  $27 \text{ in.} \times \frac{1 \text{ ft.}}{12 \text{ in.}} = \frac{27}{12} \text{ ft.} = 2\frac{7}{12} \text{ ft.} = 2 \text{ ft. } 3 \text{ in.}$

- Add: 16 ft.

  <u>+ 2 ft. 3 in.</u>

  18 ft. 3 in. Thus, the perimeter is **18 feet 3 inches**.

Finding the length of a line segment may require subtracting lengths of different units.

Example

Find the length of line segment $\overline{AB}$ in the following figure.

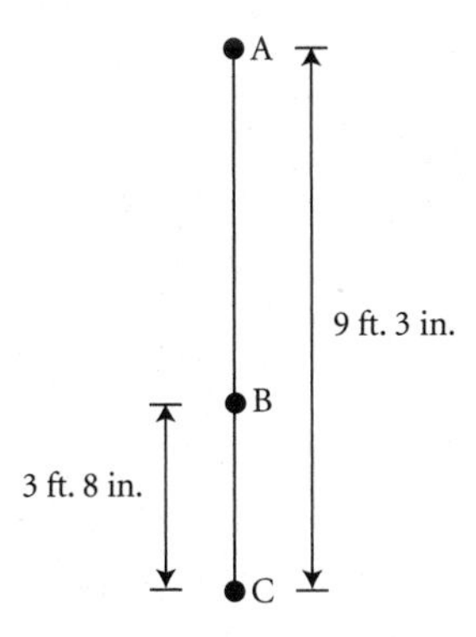

To subtract the lengths, subtract each column of length units separately, starting with the rightmost column.

9 ft. 3 in.

<u>– 3 ft. 8 in.</u>

Warning: You can't subtract 8 inches from 3 inches because 8 is larger than 3! As in regular subtraction, you must *borrow* 1 from the column on the left. However, borrowing *1 ft.* is the same as borrowing *12 inches*; adding the borrowed 12 inches to the 3 inches gives 15 inches. Thus:

$$\begin{array}{r} \overset{8}{\cancel{9}} \text{ ft. } \overset{12/15}{\cancel{3}} \text{ in.} \\ -\ 3 \text{ ft. } 8 \text{ in.} \\ \hline \mathbf{5 \text{ ft. } 7 \text{ in.}} \end{array}$$

Thus, the length of $\overline{AB}$ is **5 feet 7 inches.**

Practice

Add and simplify.

33.
$$\begin{array}{r} 20 \text{ cm } 2 \text{ mm} \\ +\ 8 \text{ cm } 5 \text{ mm} \\ \hline \end{array}$$

34.
$$\begin{array}{r} 7 \text{ km } 220 \text{ m} \\ 4 \text{ km } 180 \text{ m} \\ +\ 9 \text{ km } 770 \text{ m} \\ \hline \end{array}$$

_____ **35.** Juang Lee has one piece of molding that is 5 feet 3 inches and another piece that is 2 feet 9 inches. What is the total length of molding he has?

_____ **36.** Katherine cut the following length of fabric off a roll of popular flannel fabric on Saturday: 4 yards 2 feet, 9 yards 1 foot, 3 yards, and 5 yards 2 feet. How much of that flannel fabric did she cut in total on Saturday?

Subtract and simplify.

37.
$$\begin{array}{r} 20 \text{ m } \ \ 5 \text{ cm} \\ -\ 7 \text{ m } 32 \text{ cm} \\ \hline \end{array}$$

38.
$$\begin{array}{r} 1 \text{ mi.} \qquad\qquad \\ -\ \qquad 400 \text{ ft. } 3 \text{ in.} \\ \hline \end{array}$$

39. 2 mi. 600 yds.
– 1,700 yds.

_____ **40.** Pierre had 4 feet 1 inch ribbon and he used 2 feet 9 inches for one of the gifts he was wrapping. How much does he have remaining for his second gift?

_____ **41.** Emma has 5 yards of painter's tape and her sister Elise uses 3 yards 1 foot to tape around a window. How much tape remains?

Squares and Square Roots

Squares and square roots are used in all levels of math. You'll use them quite frequently when solving problems that involve right triangles.

To find the *square* of a number, multiply that number by itself. For example, the square of 4 is 16, because $4 \times 4 = 16$. Mathematically, this is expressed as:

$$4^2 = 16$$

4 squared equals 16

CAUTION

The notation 4^2 does NOT mean 4×2.

To find the *square root* of a number, ask yourself, "What number times itself equals the given number?" For example, the square root of 16 is 4 because $4 \times 4 = 16$. Mathematically, this is expressed as:

$$\sqrt{16} = 4$$

The square root of 16 is 4

Some square roots cannot be simplified. For example, there is no whole number that squares to 5, so just write the square root as $\sqrt{5}$.

Because certain squares and square roots tend to appear more often than others, it is sensible to memorize the most common ones in the following table.

COMMON SQUARES AND SQUARE ROOTS

SQUARES			SQUARE ROOTS		
$1^2 = 1$	$7^2 = 49$	$13^2 = 169$	$\sqrt{1} = 1$	$\sqrt{49} = 7$	$\sqrt{169} = 13$
$2^2 = 4$	$8^2 = 64$	$14^2 = 196$	$\sqrt{4} = 2$	$\sqrt{64} = 8$	$\sqrt{196} = 14$
$3^2 = 9$	$9^2 = 81$	$15^2 = 225$	$\sqrt{9} = 3$	$\sqrt{81} = 9$	$\sqrt{225} = 15$
$4^2 = 16$	$10^2 = 100$	$16^2 = 256$	$\sqrt{16} = 4$	$\sqrt{100} = 10$	$\sqrt{256} = 16$
$5^2 = 25$	$11^2 = 121$	$20^2 = 400$	$\sqrt{25} = 5$	$\sqrt{121} = 11$	$\sqrt{400} = 20$
$6^2 = 36$	$12^2 = 144$	$25^2 = 625$	$\sqrt{36} = 6$	$\sqrt{144} = 12$	$\sqrt{625} = 25$

Arithmetic Rules for Square Roots

Let's look at some rules for what *is* and what *is not* permitted when working with square roots:

$$\sqrt{a} \times \sqrt{b} = \sqrt{a \times b}$$

$$\frac{\sqrt{a}}{\sqrt{b}} = \sqrt{\frac{a}{b}}$$

These two rules show that square roots can be combined through multiplication and division by putting them under the same square root symbol and performing the given operation.

Example
What is $\sqrt{8} \times \sqrt{2}$?

Since two square roots can be multiplied and put under the same square root sign, perform this multiplication as follows:

$$\sqrt{8} \times \sqrt{2} = \sqrt{8 \times 2} = \sqrt{16} = 4$$

Example
What is $\frac{\sqrt{28}}{\sqrt{7}}$?

Since two square roots can be divided and put under the same square root sign, perform this division as follows:

$$\frac{\sqrt{28}}{\sqrt{7}} = \sqrt{\frac{28}{7}} = \sqrt{4} = 2$$

The following two rules demonstrate common mistakes that are made when working with square roots. Square roots *cannot* be combined through addition or subtraction:

$$\sqrt{a} + \sqrt{b} \neq \sqrt{a + b}$$
$$\sqrt{a} - \sqrt{b} \neq \sqrt{a - b}$$

Although it seems like square roots should be able to be combined through addition, you can see that this is not true with the following example:

$$\sqrt{100} + \sqrt{4} \neq \sqrt{104}$$
$$10 + 2 \neq \sqrt{104}$$
$$12 \neq \sqrt{104}$$

Since $12 \times 12 = 144$ (not 104), 12 is not equal to $\sqrt{104}$, and therefore, square roots cannot be combined through addition.

Practice

Use the rules on the previous page to solve these problems involving squares and square roots.

_____ **42.** $\sqrt{6} \times \sqrt{3} = ?$

_____ **43.** $\frac{\sqrt{20}}{\sqrt{5}} = ?$

_____ **44.** $\sqrt{\frac{4}{9}} = ?$

_____ **45.** $\sqrt{12} \times \sqrt{12} = ?$

_____ **46.** $\sqrt{9} + \sqrt{16} = ?$

_____ **47.** $\sqrt{9 + 16} = ?$

_____ **48.** $(3 + 4)^2 = ?$

_____ **49.** $\sqrt{7^2} = ?$

_____ **50.** $\sqrt{169 - 25} = ?$

_____ **51.** $\sqrt{169} - \sqrt{25}$?

Solving Algebraic Equations

An equation is a mathematical sentence stating that two quantities are equal. For example:

$$2x = 10 \qquad y + 5 = 8$$

The idea is to find a replacement for the unknown that will make the sentence true. That's called *solving* the equation. Thus, in the first example, $x = 5$ because $2 \times 5 = 10$. In the second example, $y = 3$ because $3 + 5 = 8$.

The general approach is to consider an equation like a balance scale, with both sides equally balanced. Simply put, whatever you do to one side, you must also do to the other side to maintain the balance. (You've already encountered this concept in working with percentages.) Thus, if you were to add 2 to the left side, you'd also have to add 2 to the right side.

Example

Apply the previous concept to solve the following equation for the unknown *n*.

$$\frac{n+2}{4} + 1 = 3$$

The goal is to rearrange the equation so *n* is isolated on one side of the equation. Begin by looking at the actions performed on *n* in the equation:

1. *n* was added to 2.
2. The sum was divided by 4.
3. That result was added to 1.

To isolate *n*, we'll have to undo these actions in *reverse order*:

3. Undo the addition of 1 by subtracting 1 from both sides of the equation:

$$\frac{n+2}{4} + 1 = 3$$
$$\underline{\quad -1 \quad -1}$$
$$\frac{n+2}{4} = 2$$

2. Undo the division by 4 by multiplying both sides by 4:

$$4 \times \frac{n+2}{4} = 2 \times 4$$
$$n + 2 = 8$$

1. Undo the addition of 2 by subtracting 2 from both sides:

$$\underline{\quad -2 \quad -2}$$

That gives us our answer:

$$n = 6$$

Notice that each action was undone by the *opposite* action:

TO UNDO THIS:	DO THE OPPOSITE:
Addition	Subtraction
Subtraction	Addition
Multiplication	Division
Division	Multiplication

Check your work! After you solve an equation, check your work by plugging the answer back into the original equation to make sure it balances. Let's see what happens when we plug 6 in for n:

$$\frac{6+2}{4} + 1 = 3\ ?$$
$$\frac{8}{4} + 1 = 3\ ?$$
$$2 + 1 = 3\ ?$$
$$3 = 3\ ✔$$

Practice

Solve each equation.

_____ **52.** $x + 3 = 7$

_____ **53.** $y - 2 = 9$

_____ **54.** $8n = 100$

_____ **55.** $\frac{m}{2} = 10$

_____ **56.** $3x - 5 = 10$

_____ **57.** $3 - 4t = 35$

_____ **58.** $\frac{3x}{4} = 9$

_____ **59.** $\frac{2n-3}{5} - 2 = 1$

TRY THIS

Do you know how tall you are? If you don't, ask a friend to measure you. Write down your height in inches using the English system. Then convert it to feet and inches (for example, 5'6"). If you're feeling ambitious, measure yourself again using the metric system. Wouldn't you like to know how many centimeters tall you are?

Next, find out how much taller or shorter you are than a friend by subtracting your heights. How much shorter are you than the ceiling of the room you're in? (You can estimate the height of the ceiling, rounding to the nearest foot.)

Answers

1. -30
2. -28
3. 3
4. 33
5. 60
6. -4
7. $-9° - 6° = -15°$
8. $-\$10{,}000 \div 5 = -\$2{,}000$
9. $\$40 - \$120 - \$35 = -\$115$
10. $\$150 - [(3 \times \$14) + (2 \times \$32) + (6 \times \$2.50)] = \$29$
11. $\$1{,}200 \times 1.5\% = \$1{,}200 \times \frac{1.5}{100} = \$18$
12. 15
13. 11
14. 6
15. 35
16. 11
17. 2
18. $\frac{25}{2}$ or $12\frac{1}{2}$
19. 4.75 ft. = 57 in.
20. 50 in. = 4 ft. 2 in.
21. 3.5 yd. = 126 in.
22. 117 in. = 3 yd. 9 in.
23. 3.2 mi. = 16,896 ft.
24. 6,336 ft. = 1.2 mi.
25. 16.8 m = 1,680 cm
26. 85,620 m
27. $1\frac{5}{6}$ ft.; 1 ft. 10 in.
28. $1\frac{1}{2}$ yds.; 1 yd. 1 ft. 6 in.
29. $1\frac{1}{2}$ mi.; 1 mi. 2,640 ft.
30. $\frac{5}{8}$ mi.
31. 34.2 cm; 34 cm 2 mm

32. 0.000165 km
33. 28 cm 7 mm
34. 21 km 170 m
35. 8 ft.
36. 22 yards 2 feet
37. 12 m 73 cm
38. 4,879 ft. 9 in.
39. 1 mi. 660 yds.
40. 1 ft. 4 in.
41. 1 yd. 2 ft.
42. $\sqrt{18}$ or $3\sqrt{2}$
43. 2
44. $\frac{2}{3}$
45. 12
46. 7
47. 5
48. 49
49. 7
50. 12
51. 8
52. $x = 4$
53. $y = 11$
54. $n = 12.5$
55. $m = 20$
56. $x = 5$
57. $t = -8$
58. $x = 12$
59. $n = 9$

GLOSSARY

angle an angle is formed when two lines meet at a point; the lines are called the *sides* of the angle, and the point is called the *vertex* of the angle

area the amount of space taken by a figure's surface; area is measured in square units

area of a circle the space a circle's surface occupies. $A = \pi r^2$

area of a quadrilateral *base* × *height*

average the sum of a list of numbers divided by the number of numbers

base the length of the side on the bottom of a polygon or quadrilateral

circle a set of points that are all the same distance from a given point, called the center

circumference the perimeter of a circle; the distance around a circle. $C = 2\pi r$ or $C = \pi d$

common denominator a denominator shared by two or more fractions

congruent two line segments of the same length

congruent angles two angles that have the same degree measure

decimal a number with digits only to the right of the decimal point, such as 0.17; a decimal's value is always less than one

decimal value when the numerator is divided by the denominator in a fraction, the result is the decimal value of the fraction; for example, the decimal value of $\frac{3}{4}$ is 0.75

denominator the bottom number in a fraction: 2 is the denominator in $\frac{1}{2}$

diameter the length of a line that passes across a circle through the center; the diameter is twice the size of the *radius*; and the symbol *d* is used for the diameter

difference the number left when one number has been subtracted from another

divisible by a number is divisible by a second number if that second number divides *evenly* into the original number: 10 is divisible by 5 ($10 \div 5 = 2$, with no remainder), however, 10 is not divisible by 3 (see *multiple of*)

equilateral triangle a triangle with 3 congruent angles each measuring 60° and 3 congruent sides

even integer integers that are divisible by 2, such as 24, 22, 0, 2, 4, and so on (see *integer*)

fraction a part of a whole; the two numbers that compose a fraction are called the *numerator* and *denominator*, and decimals, percents, ratios, and proportions are also fractions.

height the length of a perpendicular line drawn from the base to the side opposite it; *altitude*

hypotenuse the longest side of the triangle

improper fraction a fraction whose numerator is greater than its denominator; an example is $\frac{22}{5}$

integer a number along the number line, such as 23, 22, 21, 0, 1, 2, 3, etc.; integers include whole numbers and their negatives (see *whole number*)

is equals (=)

isosceles triangle a triangle with 2 congruent angles, called base angles; the third angle is the *vertex angle*, the sides opposite the base angles are congruent, and an equilateral triangle is also isosceles.

least common denominator the smallest number into which two or more denominators (in two or more fractions) can divide evenly

legs the two shorter sides of a triangle, opposite the *hypotenuse*

line an infinite number of points that extend endlessly in both directions

line segment a section of a line with two endpoints

mean (See *average*)

measures of central tendency numbers that researchers use to represent data; averages, such as *mean*, *median*, and *mode*

median the term in the middle position of an arranged data set. It is the average of the middle two terms if there is an even number of numbers in the list.

midpoint a point on a line segment that divides it into two line segments of equal length, which are called *congruent lines*

mixed decimal a decimal in which numbers appear on both sides of the decimal point, such as 6.17; a mixed decimal's value is always greater than 1

mixed number a number involving a whole part and a proper fraction; an example is $3\frac{1}{5}$

mode the most frequently occurring member of a list of numbers

multiple of a number is a multiple of a second number if that second number can be multiplied by an integer to get the original number: 10 is a multiple of 5 ($10 = 5 \times 2$); however, 10 is not a multiple of 3 (see *divisible by*)

negative number a number that is less than zero, such as –1, –18.6, $-\frac{1}{4}$

numerator the top part of a fraction: 1 is the numerator in $\frac{1}{2}$

obtuse angle an angle measuring more then 90° and less than 180°

odd integer an integer that is not divisible by 2, such as –5, –3, –1, 1, 3, and so on

of multiply (×)

parallel lines lines that lie in the same plane and do not cross or intersect at any point

percent part of a whole; x% means $\frac{x}{100}$

perimeter the distance around a polygon; the word *perimeter* is derived from *peri*, which means *around*, and *meter*, which means *measure*

perimeter of a quadrilateral the sum of all four sides of a quadrilateral

perpendicular lines lines that lie in the same plane and cross to form four right angles

polygon a closed, planar (flat) figure formed by three or more connected line segments that do not cross each other. The three most common polygons are triangles, squares, and rectangles.

positive number a number that is greater than zero, such as 2, 42, $\frac{1}{2}$, 4.63

probability the number of favorable outcomes that could occur divided by the total number of possible outcomes; probability is expressed as a ratio

product the answer of a multiplication problem

proper fraction a fraction in which the top number is less than the bottom number (numerator is lesser than the denominator)

proportion two ratios are equal to each other

Pythagorean theorem $a^2 + b^2 + c^2$ (c refers to the hypotenuse)

quadrilateral a four-sided polygon; the three most common quadrilaterals are rectangles, squares, and parallelograms

quotient the answer you get when you divide: 10 divided by 5 is 2; the quotient is 2

radius the distance from the center of the circle to any point on the circle itself; the symbol r is used for the radius

ratio a comparison of two numbers: *nine out of ten* is a ratio

real number any number you can think of, such as 17, –5, $\frac{1}{2}$, –23.6, 3.4329, 0; real numbers include the integers, fractions, and decimals (see *integer*)

reduced fraction a fraction written in lowest terms; when you do arithmetic with fractions, it is best to always reduce your answer to lowest terms

remainder the number left over after division: 11 divided by 2 is 5, with a remainder of 1

repeating decimal a decimal that results from a fraction that cannot be divided evenly: $\frac{2}{3}$ becomes a repeating decimal because it divides as 0.66666

right angle an angle measuring exactly 90°

right triangle a triangle with a 90° angle; a right triangle may be isosceles or scalene

rounding estimating the value of a decimal using fewer digits

scalene triangle a triangle with no congruent sides and no congruent angles

straight angle an angle measuring exactly 180°

sum the resulting number when two numbers are added together

triangle a polygon with three sides. Triangles can be classified by the size of their angles and sides.

unknown a letter of the alphabet that is used to represent an unknown number; *variable*

variable a letter of the alphabet that is used to represent an unknown number; *unknown*

vertical angles two angles that are opposite each other when two lines cross

whole number numbers you can count on your fingers, such as 1, 2, 3, and so on; all whole numbers are positive

APPENDIX I: INTERVIEWING FOR SUCCESS

The job search process can be a time-consuming one. After you have found the right job opportunities to apply for, and then submitted a well-written cover letter and a finely tuned resume to capture the potential employer's attention, then it is *your* responsibility to impress the person who invites you in for a job interview. Only after a successful interview might you receive a job offer.

This appendix introduces you to the different types of interviews used in the professional world and provides you with essential guidelines to follow as you prepare for great interviews.

Throughout each of your interviews, there will probably be a lot on your mind. What will be on the interviewer's mind, however, are these questions:

- Will this person be successful in the job if he or she is hired?
- Will this person be an asset to the company?

- Is this person worth the salary the company would be paying him or her?

Everything you do during an interview should help answer these questions in a positive way.

If you are nervous, don't worry. This chapter is designed to help you prepare for different job interview styles, conduct yourself professionally during any interview situation, and evaluate any job offer you receive.

Creating a Great Impression

Being invited to participate in a job interview is a positive indication that your cover letter and resume have done their job. The experience and skills you displayed on paper have gotten a company interested in the possibility of hiring you. Now the employer wants to meet you in person, get to know you better, and learn more about your qualifications.

A job interview is your opportunity to sell yourself directly to a potential employer. From the moment you step into an employer's office, everything about you will be evaluated, including all the following:

- appearance (personal grooming)
- attitude
- outfit
- body language
- personality
- communication skills
- level of preparation

The job interview is your big chance to impress the interviewer with everything you have to offer. It is your opportunity to set yourself apart from the other applicants, demonstrate that you are qualified to fill the available job opening, and show how excited you are about the prospect of working for that employer.

Scheduling the Interview

Ideally, shortly after you send your resume and cover letter to a potential employer, your telephone will ring (or a message will pop up in your inbox) and you will be invited in for an interview. When you are on the phone with a representative from the company, always act professionally and remember to obtain the information you need from the person with whom you are speaking. If your prospective employer reaches out via e-mail, make sure you are just as professional and poised as you would be over the phone.

There are four important rules for scheduling your interview:

1. **Be easy to schedule.** The interviewer, or his or her assistant, will likely start off the scheduling chat with a date. Try to be flexible. If she says "next Thursday at 3:00 P.M.," and that's when your weekly staff meeting takes place, reply with a better time for you, such as "next Thursday at lunchtime." However, watch how you say it. Don't be too specific, such as, "next Thursday at 12:15"—keep "lunchtime" open to his or her interpretation. Likewise, if before 9:00 A.M. is the best time for you, go ahead and say so, but if that time slot doesn't work for the interviewer, offer another one, such as during your lunch hour or after 5:00 P.M.
2. **Keep your scheduling details organized.** A person can be so excited, thrilled, or nervous about getting called for an interview that he or she can forget or misunderstand the details. First, pause and collect yourself. Then, thank the person on the other end of the phone line for calling, and ask for a moment to grab a pen and paper along with your planner.

3. **Remember your manners.** You never know who will have a say in the hiring decision, so why risk your future by using bad manners? Often, when a potential employer or human resources representative calls you to talk about setting up an interview, he or she is getting a sense of your personality over the phone.
4. **Schedule your interview with plenty of time to prepare.** Always avoid scheduling an interview for the same day that you are called to schedule an interview. Allow yourself at least one day, or preferably two, to prepare and do your research.

TIP

Here is a list of information you need to get during the initial interview scheduling call:

- the name of the person conducting the interview (along with his or her title)
- the exact position for which you are interviewing
- the location of the interview
- directions to the location (or make a note to yourself to Google directions ahead of time)
- the name of the person to ask for at the interview location, as well as that person's phone number and extension
- what additional materials, if any, to bring (such as a portfolio, samples, or reel)

At the end of the conversation, it is essential to do two things:

- confirm date, time, place, and materials to bring
- say "thank you"

Dos and Don'ts

Here are some examples of the right and wrong way to schedule an interview with a potential employer.

Scheduling the Right Way and Wrong Way: Sample 1

Wrong: *Yeah, I can come in sometime next week, I guess.*

Wrong: *Who? Oh, right. The marketing job. Uh, I'm not sure what my day looks like on Tuesday. Where's my iPhone? Let me call you back. . . .*

Wrong: *Wednesday? I'm supposed to go out of town tomorrow—can I get back to you in like a week or two?*

Wrong: *Smrrrring. Bhaah . . .[inaudible or mumbled responses]*

Right: *Thank you for calling, Ms. Peterson. I am very interested in meeting with you to discuss the open position in the marketing department. I have several times available this week—are mornings or afternoons more convenient for you?*

Scheduling the Right Way and Wrong Way: Sample 2

Wrong: *Uh, okay, see you then. [Click.]*

Wrong: *Thanks. Bye.*

Wrong: *I hope to see you then. Ciao.*

Right: *Next Thursday, November 14th at 12:30 sounds perfect. I will see you at Milford Corporate Park in Human Resources reception, and I will bring my portfolio. Thanks again. Goodbye.*

Scheduling the Right Way and Wrong Way: Sample 3

Wrong: *Um, 3:00 P.M. is no good.*

Wrong: *I can't make it.*

Wrong: *Jeez, that's my kickboxing class time, and I always spend my lunch hour in that class on Mondays.*

Right: *Sorry, I have a staff meeting at that time. How does a bit earlier in the day work for you? I can come in around lunchtime.*

CAUTION

Keep in mind that the scheduling phone call or e-mail is often a part of the screening process. If your attitude was rude, shy, or negative in any way, it was probably noted. An overly casual or blasé e-mail will not get your interviewer excited about meeting with you, but your positive qualities will also be noted, so always be professional and respond in a timely manner.

Pre-Interview Research

The first step when preparing for any interview is to do research. Always enter the interview knowing as much as possible about the employer and the related industry. Specifically, here are some of the details about which you want up-to-date and accurate information:

- **The job for which you are applying.** Know exactly what position you are hoping to fill, what the requirements are for that position, what skills/training are required, and what are the company's needs.
- **The industry in which you will be working.** How big is the industry? What are the biggest companies in the industry? What are the challenges facing the industry as a whole? Is the industry growing? Knowing this information, even though it may not be asked during the interview, will help you feel comfortable and confident.
- **The company with which you will be interviewing.** What is the company's history? What does the company do or sell? What sets it apart from the competition? What are the strengths and weaknesses of the company? Who would be

your boss? What challenges is the company facing in the future?

- **The person who will be interviewing you.** Find out the title of the person interviewing you, as well as his or her responsibilities within the company. Start thinking about what questions you will ask this person.

TIPS

Research resources:

- Visit the employer's website, in addition to the websites of the employer's main competition.
- Read company-issued press releases, company newsletters, and industry-specific magazines or websites.
- Use social media—LinkedIn, Facebook, Twitter—to search for companies and people you may know.
- Check the chamber of commerce in your city or town.
- Speak directly with people who already work for the company or in that industry.
- For a civil service position: Obtain information from the agency's websites or contact the department directly.

Rehearsing Your Material

Even if you can't perfectly predict what an interviewer will ask you, it's a safe bet he or she will want to know about your previous job experience, how the responsibilities you've had in the past will translate to the position for which you're interviewing, a time you excelled at work, a time you overcame a challenge or conflict, and what you can bring to the table at this particular organization. The research you've done on this job and this company will help with some of the answers, but it's up to you to make the most out of your personal history.

Practice with a friend or family member—ask her or him to ask you an open-ended question from the preceding list, and see how well you can tell an anecdote that demonstrates an ability you have or a skill you acquired. Try to identify at least one specific story from each job that you can use to illustrate why you're a great fit for this company. Use your resume as a starting point to jog your memory if you need to. Your interviewer will want to hear you think on your feet in order to assess your communication skills, and you can show off both your memory and your self-awareness by having relevant anecdotes already drafted and rehearsed.

Choose the Right Thing to Wear

Your appearance greatly affects an employer's first impression of you. Never wait until the last minute to choose your interview outfit, accessories, or hairstyle. What should you wear to an interview? A lot depends on the company's established dress code and culture. Your main goal is to look professional. No matter who you are, where you live, what job you are applying for, or what type of company you are visiting, your outfit should be clean, well-tailored, flattering, and wrinkle free.

What to Wear—Men

- a well-tailored, clean, and pressed suit in conservative, dark shades of navy blue, gray, or brown
- in a non-corporate environment, an acceptable alternative for a suit is a (less formal) sport jacket/blazer and dress slacks
- a pressed white or light colored, long-sleeved, cotton dress shirt

- a tie that coordinates with your suit, avoiding wild colors and patterns
- dark socks that coordinate with your suit and dress shoes
- polished, plain black or brown leather dress shoes

What to Wear—Women

Women can be a bit more creative in their wardrobe selection, as long as it fits within the company's dress code. You can look equally professional in a tailored dress or a blouse with a skirt or dress slacks. Hosiery, depending on the season and region, is an important consideration. In addition:

- polished, plain, sensible pumps, or low-heeled dress shoes
- natural-looking makeup
- simple and understated jewelry that complements your outfit without attracting attention

What to Avoid—Men

- jewelry, other than a watch or your wedding band
- baseball caps or other hats

What to Avoid—Women

- low-cut necklines, sleeveless tops, and sheer fabrics
- miniskirts
- loud prints and patterns
- open-toed shoes and spike heels
- dramatic makeup
- excessive or flashy jewelry, overtly religious symbols

General

- Make sure that you wash and neatly style your hair.
- Be sure to wear deodorant/antiperspirant.
- Clean and trim your nails.

- Avoid fragrance of any kind—scents can cause allergic reactions in others, or may be considered unappealing by your interviewer.
- Body piercings: Wear small, simple earrings; no dangle earrings.
- Tattoos: Your pre-interview research will tell you whether the office culture is likely to be accepting of tattoos; consider covering them if the workplace seems to be conservative.

What to Bring

When you are stressed about your interview, it's easy to forget to bring something with you. Here is a list of everything you'll likely need:

- Several extra copies of your resume, letters of recommendation, and your list of references
- Your daily planner, phone, or tablet (so you can easily schedule additional appointments)
- Folder containing company research materials, a pad, and two working pens
- Any additional materials (such as writing samples, portfolios, clips) requested by the interviewer or included in your application

All these items will fit into a briefcase or portfolio. Write down the company's name, interviewer's name, address, telephone number, and directions to the location of the interview the night before. This way, there's no chance of losing or misplacing vital information.

Common Types of Interviews

Being invited to participate in a job interview is a positive sign that your cover letter and resume have done their job. Now the employer wants to meet you in person and learn more about your qualifications.

Potential employers can use many interview styles to get to know you better. Some interviews are done in a private office, on a one-to-one basis. Some interviews are done over the telephone or Skype, or over lunch or dinner at a restaurant. Following are descriptions of the most common types of interviews.

One-on-One Interview

This is the most common type of interview. It involves two people—you and the interviewer. Most likely, you will be sitting opposite each other in an office or conference room while participating in a two-way conversation. However, these interviews can sometimes happen offsite (usually over a meal in a restaurant), or even over the telephone or Skype (usually because you live out of town).

The Human Resources Screening Interview

Often, a representative from an employer's human resources department will interview you before you meet with the hiring manager. The screening interview ensures that you are right for the job. Inappropriate candidates (people who are obviously unprepared, unprofessional, etc.) are screened out. This first step saves the hiring manager's time.

Group Interview

While this may seem unusual, it is actually a common format, especially for big companies doing campus recruiting. If you are asked to participate in a group interview alongside several other applicants, your main priority is to make the most out of the attention you get. When it is your turn to respond to a question, it is your opportunity to make yourself stand out from the other applicants. Describe your marketable attributes clearly and concisely, and display your enthusiasm for the job and the company. In this situation, you will have less time to win over the interviewer, but at the same time, you can size up the competition. Never be rude or interrupt fellow applicants.

Panel Interview

During a panel interview, you will meet with several people at the same time. This type of interview simulates a business meeting at which you are the presenter. Members of the panel may be individuals with whom you would interact on the job or individuals designated by the company as a hiring committee or employee search group. The panel may include your potential supervisor and/or a human resources representative.

Being interviewed by a panel of people adds a bit of a challenge to your objective (which is to impress the interviewer). You now need to impress two, three, or more people at once. In addition to intelligently answering the questions posed to you and asking insightful questions, it is vital that you maintain eye contact and develop a rapport with each of them. Do not allow yourself to become intimidated in this situation. The interviewers know a panel interview format adds pressure, and they want to see how you will react. Try to give the panelists equal time. Providing thorough answers to each question ensures that you've given respectful, thoughtful answers to each member. Remember to be flexible and to demonstrate that you can think on your feet. If you go into the interview having done your research and are totally prepared, it will not matter whether one person or five people are conducting the interview.

The Second Interview

Sometimes, the interview process is a long one and can be spread out over different days. If you are invited back for a second interview, expect the interviewer to ask you more detailed and specific questions that directly relate to the job you are applying to fill. If you are invited for a second or third interview, you know the employer is interested in you. Your job now is to say everything in your power to convince him or her to hire you.

Do not be surprised if you are introduced to other executives within the company during and after the second and third interviews. Also, the formats of subsequent interviews might change. The second interview, for example, might be held over lunch at a restaurant or be conducted as a panel interview.

Whatever style or format your second and third interviews are, you must be prepared. Do not get lazy and think of the additional interviews as a mere formality. As the interview process goes on, the importance of each interview actually increases, so do your best every time.

Top 25 Interview Mistakes

We can't emphasize these facts enough, so here they are, ready for you to learn. With this list, you'll find out which actions are inappropriate, what comments to avoid, and what not to forget. You will be glad you did—it could mean the difference between being hired and being passed over.

1. **Showing up late.** Be sure to allow yourself ample time to get to your interview. You must factor in unexpected circumstances, such as train delays or heavy traffic. If you know you're going to be late, call to let your interviewer know.
2. **Being unprepared to describe your experiences.** An interview is a test—and you should never walk into a test unprepared. Take some time to prepare your rehearsed responses, and think about how you will handle the questions your interviewer might ask. Practice describing your experiences aloud or conduct practice interviews with a partner.

3. **Answering questions with only a *yes* or *no*.** Your interviewer needs to get to know you, and he or she will be unable to do that if you don't volunteer information about yourself. Be sure to support your answers with examples.
4. **Fidgeting.** If you are tapping your foot, playing with a bracelet on your wrist, or constantly shifting in your seat, you won't look professional. And if you don't look professional, you won't be hired.
5. **Speaking too quickly.** You may want to get in a lot of information, but you don't want to speak so fast that your interviewer can't understand you. Take a deep breath before you begin answering questions and slow yourself down. Conduct practice interviews with a friend to make sure that your speaking voice is steady and even.
6. **Avoiding eye contact.** If you avoid making eye contact, you will be unable to establish a personal connection with your interviewer. You should be attentive and engaged in what your interviewer is saying.
7. **Not researching the company.** Q: "What do you know about our firm?" A: "Uh . . . not much." Answers like this will not get you hired.
8. **Lying.** Don't lie about or embellish your job experiences or academic record. Your interviewer is going to check these things out. If an interviewer catches you lying, you won't be hired. If your employer finds out about your misrepresentation after you've been hired, you will be fired—it's as simple as that.
9. **Not answering the question asked.** You want to highlight your experiences in the interview, but you should be careful to always answer the question being asked. Don't be so intent on launching into a great story about you that you avoid the question altogether—your interviewer will notice.
10. **Revealing too much.** Your interviewer is neither your best friend nor your therapist. She wants to learn about

the skills and qualities you will bring to a job. She does not want to hear about your personal life or problems.

11. **Not "selling" yourself when you answer questions.** You should answer questions in a way that brings out the qualities that will serve you on the job. If you are asked how your best friend would describe you, say something like, "I think my best friend would describe me as loyal and dependable. People always know that they can count on me."
12. **Speaking poorly of, or belittling, past job experiences.** Disparaging other employers or jobs will make you sound unprofessional, negative, and hostile. And it will make the interviewer wonder what you would say about his or her company to others. Try to focus on what you learned from other jobs.
13. **Dressing too casually.** Your interviewer wants to hire a responsible professional. Make sure you look like one.
14. **Not asking any questions about the company.** By asking some good questions, you will prove that you are very interested in the job—and that you were motivated enough to research the position and the company.
15. **Not thanking the interviewer at the end of the interview.** In the business world, a little courtesy goes a long way. Your interviewer will appreciate and notice your good manners.
16. **Forgetting to send a thank-you note.** Demonstrate your professionalism and courtesy by sending a note. You will also be more likely to stand out in your interviewer's mind if he or she has a reminder of the interview.
17. **Forgetting to bring extra resumes to the interview.** You may be asked for another copy of your resume, and you may have to submit an extra copy with any forms you have to fill out. Make sure that you are prepared.
18. **Neglecting to prepare a list of references.** Type up your references (with contact information) for your interview-

er. He or she will not be interested in taking down all the names and numbers by hand, and it will be an inconvenience if you have to send the information at a later date.

19. **Forgetting the interviewer's name.** You should always bring a note pad (preferably in a professional leather portfolio) to an interview. Write down the interviewer's name if you think you won't be able to remember it. Thank the interviewer by name at the end of the interview.
20. **Going to an interview on an empty stomach.** You will feel more alert if you've had a nutritious meal, and you won't get hungry if the interview ends up lasting much longer than you had anticipated. And, of course, you won't have to worry about your stomach rumbling in the middle of a question.
21. **Using slang**. Nothing makes you sound more unprofessional than peppering your speech with *like* and *y'know*. If you can't speak like a professional, your interviewer will question whether he or she can trust you to interact with clients or supervisors.
22. **Chewing gum, eating, or smoking.** These are obvious no-nos.
23. **Answering your cell phone.** Turn off your cell phone before you get to the interview.
24. **Interrupting the interviewer or talking excessively.** Don't ramble or go off on tangents. You want to showcase yourself and give the interviewer a good sense of your accomplishments, but make sure you don't cut the interviewer off or preclude him or her from asking questions. He or she has limited time to speak with you.
25. **Freezing up.** Relax! It's only an interview. If you are well prepared, you should feel confident and stress-free. Smile and be yourself.

Just Before the Interview

Okay, you are just about ready to participate in any job interview situation. When you get to the interview (15 minutes early), do these final things:

- Visit the restroom (if for no other reason than to check your hair, tie knot, or lipstick in the mirror).
- Glance at your notes to refresh your memory.
- Scan a corporate publication, newspaper, or trade magazine if offered in the reception area.
- Smile and be polite to everyone you encounter.
- Practice deep, calming breaths.
- Visualize yourself having a great interview.

During the Interview

Although every interview you participate in will be unique, most interviews have five stages:

1. Introductions
2. Small talk
3. Exchanging information
4. Summarizing
5. Closing

Practice each stage at home, recording yourself on video or enlisting the help of a friend who can give you feedback on each stage. Be sensitive to your interviewer's transitions from one section to another; if it seems like they're beginning to wrap up, don't try to prolong the interview unnecessarily.

Take some time to review these guidelines, which are essential for succeeding on any interview.

- Get a good night's sleep before the interview so that you are awake and alert. You want to look and feel rested.
- The morning of your interview, read the local news and/or watch a morning news program. Be aware of the day's news events and be able to discuss them with the interviewer. Many interviewers like to start an interview with general chitchat. You want to appear knowledgeable about what is happening in the world around you.
- Before you enter the building for an interview, turn off your cell phone. In addition, turn off any alarms you have set. It is unacceptable to interrupt any interview with annoying sounds.
- Arrive at the interview alone. This may seem obvious, but even if a parent, significant other, or friend drove you to the interview or commuted with you for moral support, leave him or her outside. You want to appear confident, not codependent.
- Arrive early for your interview, no matter what. Ideally, you want to arrive about 10 to 15 minutes early and check in with the receptionist. Don't arrive too early, though. Interviewers often have back-to-back meetings or other priorities. Arriving too early can be misinterpreted as not respecting the interviewer's carefully arranged schedule for the day.
- Keep in mind that you are being evaluated on every move you make. Often, even the receptionists or assistants have input on the hiring process, so always act professionally. Be polite, courteous, and friendly to everyone you encounter.
- When you are introduced to the interviewer, stand up, smile, make direct eye contact, and shake hands. Refer to the interviewer formally, as Mr./Ms./Dr. (insert last name).
- Throughout the interview, always try to maintain as much eye contact as possible and avoid fidgeting. The interviewer is probably trained to read your body language. Avoiding eye contact and touching your face are often signs of someone who is not being truthful.

- Never smoke or chew gum in an interview.
- You may be offered a drink at the start of the interview. Accept whatever is offered. To avoid caffeine, consider asking for water. If you are at a lunch or dinner meeting and the employer offers you an alcoholic beverage, decline, but ask for something nonalcoholic.
- During your interview, be prepared to make small talk—about the weather, how a local sports team is performing, or a major news story.
- Even if the interviewer takes a casual approach to the interview, you are still being evaluated, so never lose focus. Pay attention to the conversation, keep smiling, maintain eye contact, answer all questions openly and honestly, and use complete sentences. Never answer questions with a one-word answer of yes or no.
- Always think before you speak. Your interviewer asks every question—even if it seems to have no relevance to anything—for a reason. Some employers like to see how you will react to bizarre questions, and to see whether you can think on your feet. Often, what you provide as an answer to these off-the-wall questions is not as important as how you answer them.
- If you do not know the answer to a specific question, never lie or make up an answer. Never say, "I don't know." When you feel stuck, pause. Take a breath. Say "Hm, let me think about that for a moment" and proceed when you feel calm and prepared. What may seem like an eternity to you will come across as a contemplative, thought-gathering moment to the interviewer.
- At your initial interview, you should not ask about salary, benefits, or vacation time. You can bring up these topics once the interviewer offers you the job or expresses a very strong interest in hiring you—not before.
- During the later part of your interviews, be sure to affirm that you are interested in the job for which you are applying. Explain exactly why you want the job, what you can offer to

the company, and why you are the best candidate to fill the position.

- Follow up the interview by sending a thank-you note or e-mail to the interviewer within 24 hours.

After the Interview

Send a Thank-You Note

Sending a short, personalized thank you note within 24 hours after an interview is an absolute must. Send it to the interviewer via e-mail or as a legibly handwritten or typed note. In your note, make a specific reference to something you discussed during the interview (to refresh the interviewer's mind about who you are), and explain in a sentence or two how excited you are about the possibility of working for the company.

The note itself can take a couple of different forms, depending on the industry and your personal taste:

- **The typed letter.** Using the same letterhead and paper that you created for your resume, type out a brief, professional letter on a computer. Print it out, and send it in a matching envelope. Don't forget the stamp!
- **The handwritten note.** Some people prefer the personal touch. If you have personalized stationery with your name or initials on it, go ahead and write out a brief note. Don't make the mistake of using casual language in the note: Be specific, be direct, and be professional. Also, it is a good idea to gather your thoughts and edit on scrap paper, then copy the finished note onto the good stock. Many interviewers appreciate the personal touch; however, if you are interviewing in a conservative industry, use the typed letter instead.

Sample Interview Follow-Up and Thank-You

Linda uses her letter to outline how her accomplishments meet or exceed the ideal candidate qualifications discussed during her interview.

Linda Jameson
2123 Lincoln Avenue, Apt 3D, Minneapolis, MN 55555, 555-555-1234
linda.jameson@yourdomain.com

May 14, 2010

William Sanders
Prudential Financial Services
600 U.S. 169
St. Louis Park, MN 55555

Dear Mr. Sanders:

Thank you very much for the time and opportunity to meet with you this past Thursday. Our discussion was enlightening and deepened my interest in the Prudential Financial team. I remain confident that my qualifications are a strong match for your needs, and I hope to be among those in consideration for the position of Financial Services Associate.

My Bachelor of Arts degree in Finance from Northeastern University, combined with my three years of experience at MetLife, has provided me with the right background to become an asset to Prudential. During my time at MetLife, I have collaborated with my team members to achieve company goals, maintain relationships with a broad client base, and continue to expand my capabilities by enrolling in relevant educational programs in my personal time. Joining the Prudential team will allow me to grow and be challenged in a successful environment.

Thank you, once again, for your time. I will remain available for further interviews as needed. I look forward to hearing from you again soon.

Sincerely,

Linda Jameson

Follow Up

After the interview and depending on the "next steps" discussed at the close of the interview, you should follow up. Generally, the guideline is to follow up three to five business days after the interview, either by telephone or e-mail.

To set yourself apart from other candidates, send a thank-you note via e-mail immediately following the interview, and also send a note on your stationery or bonded resume-quality paper via standard mail.

APPENDIX II: WORKPLACE DOS AND DON'TS

So you've received the job offer, you've filled out all the paperwork, and read through the company orientation manual. Now what?

Your First Day in the Office

Your supervisor or a new colleague will most likely show you around—depending on the size of the company, this may be as simple as pointing out the communal kitchen, the bathroom, and the network printers, or as elaborate as a tour of multiple floors of an office building to learn where various departments are. Although either tour will probably include a visit to the office supply cabinet, make sure you bring something to write with and something to write on to your first day. If you're invited to attend meetings right away, sit back and observe; it's usually better to ask questions to your immediate supervisor or office neighbor to catch up later instead of disrupting an ongoing discussion. Draw yourself a seating chart to record your new coworkers' names and where they sat—it will help you to put names with faces—and jotting down some notes of what was said will begin to get you up to speed.

Joining a new company often means learning what systems are already in progress—who do you report to? How are you given new assignments? Do you get to determine what you work on and when, or will your supervisor be managing your day-to-day tasks? If you don't already know, be sure you find out during your first week. You can't meet or exceed expectations if you don't know what they are first!

DOs

- **Ask questions!**
- **Take initiative:** This doesn't mean going rogue or being resistant to existing company practices, but you can demonstrate that you're an independent thinker by coming up with your own way to complete an assignment, and running it by your supervisor to get his approval. If your superior wants to redirect you, be receptive to the feedback, but perhaps she'll appreciate that you're already thinking of ways to innovate.
- **Get to know your neighbors:** Don't be distracting or monopolize people's time, but set a goal of introducing yourself to one coworker a day until you know at least everyone on your team, or anyone with whom your department interacts regularly. Even just a few moments of chatting as you finish a first cup of coffee can help you develop working relationships with your colleagues.
- **Volunteer for projects:** Obviously this should be cleared with your supervisor, but a great way to expand your responsibilities in a new position is to be willing to take on

new challenges. If you're in a meeting and someone higher up the food chain is looking for someone to lead a new project or supervise the execution of a new initiative, consider whether your workload could accommodate an addition. It's almost always better to be the person who says "Yes, I can handle that for you—anything else?" than the person who sits silently while an opportunity passes by.

DON'Ts

- **Complain:** Even if your old employer gave out free coffee and omelets every morning, and all your new office seems to have is stale animal crackers, you should approach your first weeks on the job with a continuation of your best interview behavior. Don't let your reputation become that of somebody who finds fault with everything, who gripes instead of saying good morning, or who is convinced the grass was greener on the other side of the fence.
- **Act helpless:** If you're really and truly stuck with something, you can always ask for help instead of wasting time struggling under the radar. But with run-of-the-mill IT issues, small-scale office-related needs, or new software to learn, try to cultivate a sense of self-sufficiency where you can. A needy employee can become a distraction to coworkers or signal the boss that you're not ready for more responsibility.
- **Get too comfortable too quickly:** Many workplaces allow employees to customize their workspaces in some ways; others prefer to keep shared spaces uncluttered and uniform. Be aware of the prevailing office culture and do your best not to disrupt it by being too loud, letting your personal effects creep onto someone else's desk, or bringing in too many photos or knickknacks from home. There's always time to bring more of your personality into the office once you've

established yourself as a professional first, and a cat or dog or sports enthusiast second.

- **Violate or mock HR policies:** Whether your orientation was a quick spin around the office complex or a more formal companywide presentation, the guidelines presented by your HR rep were given to you for a reason. You can demonstrate your professionalism by taking them seriously, from the basics, like adhering to a dress code, to the legal standards of conduct, like refraining from making personal comments about your coworkers.

ADDITIONAL ONLINE PRACTICE

Using the codes below, you'll be able to log in and access additional online practice materials!

Your free online practice access codes are:

FVEM6DJFXQ48K50360E1
FVE43HFYT08M0W1GKPRH
FVEFCW2O01O4T63EBKMD
FVELDKND48W6423175KL
FVEX0DT7MUO1AEKNS5PP
FVEYUU15QNK4P521C21C
FVE456Q7YNDSUFN46OAD
FVE1VDU33YQX45YK156G
FVE4241RS24HQOU4LB2R
FVE41EA41WI2FMQJBN2B
FVE2D3GV64135BOLRQ5R

Follow these simple steps to redeem your codes:

- Go to **www.learningexpresshub.com/affiliate** and have your access codes handy.

If you're a new user:

- Click the **New user? Register here** button and complete the registration form to create your account and access your products.
- Be sure to enter your unique access codes only once. If you have multiple access codes, you can enter them all—just use a comma to separate each codes.
- The next time you visit, simply click the **Returning user? Sign in** button and enter your username and password.

- Do not re-enter previously redeemed access codes. Any products you previously accessed are saved in the **My Account** section on the site. Entering a previously redeemed access codes will result in an error message.

If you're a returning user:

- Click the **Returning user? Sign in** button, enter your username and password, and click **Sign In**.
- You will automatically be brought to the **My Account** page to access your products.

- Do not re-enter previously redeemed access codes. Any products you previously accessed are saved in the **My Account** section on the site. Entering a previously redeemed access codes will result in an error message.

If you're a returning user with a new access codes:

- Click the **Returning user? Sign in** button, enter your username, password, and new access codes, and click **Sign In**.
- If you have multiple access codes, you can enter them all—just use a comma to separate each codes.

- Do not re-enter previously redeemed access codes. Any products you previously accessed are saved in the **My Account** section on the site. Entering a previously redeemed access codes will result in an error message.

If you have any questions, please contact LearningExpress Customer Support at LXHub@LearningExpressHub.com. All inquiries will be responded to within a 24-hour period during our normal business hours: 9:00 A.M.–5:00 P.M. Eastern Time. Thank you!

NOTES

NOTES

NOTES

NOTES

NOTES

NOTES